Vitamin K₂ and the Calcium Paradox

How a Little-Known Vitamin Could Save Your Life

Kate Rhéaume-Bleue

Collins

Vitamin K$_2$ and the Calcium Paradox
Copyright © 2012 by Kate Rhéaume-Bleue.

Published by Collins, an imprint of HarperCollins Publishers Ltd

Originally published by John Wiley & Sons Canada, Ltd.
in both print and EPUB editions: 2012

First published by Collins, an imprint of HarperCollins Publishers Ltd,
in an EPUB edition and in this trade paperback edition: 2013

HarperCollins books may be purchased for educational, business, or sales
promotional use through our Special Markets Department.

HarperCollins Publishers Ltd
2 Bloor Street East, 20th Floor
Toronto, Ontario, Canada
M4W 1A8

www.harpercollins.ca

Library and Archives Canada Cataloguing in Publication
information is available upon request

ISBN 978-1-44342-807-1

Printed and bound in U.S.A.
23 24 25 26 27 LBC 26 25 24 23 22

To Sterling, I hope by the time you are grown up, this will all be common knowledge.

Contents

Acknowledgments

First of all I'd like to thank my amazing literary agent, Rick Broadhead. Rick immediately recognized the value of my idea, helped me shape the proposal, found a home for my book at Wiley and supported me in every step of the publishing and marketing process. Thanks, Rick.

Thank you to my family for their support and patience throughout the writing process and beyond. I'm especially grateful to my husband, Chris, for his encouragement and analogies, and to my in-laws, Linda and David, for frequent babysitting and dog walking. To my own Mum and Dad, thanks for instilling in me a sense of curiosity, an appreciation of language and determination. Also to my sister, Robin, thanks for being in my corner.

Thank you to my many dear friends who supported me along the way—whether you knew it or not. I'm particularly appreciative of you, Rahima, Paula, Jenny, Cher, Lynn, Lara, Lisa and Joyce.

Just in case I don't get a chance to say it elsewhere, I'm deeply appreciative of every member of my "Factors Family"—you know who you are—for support and inspiration. I'm especially thankful to Joanne Aldridge. Jo, thanks for keeping me organized, putting out fires and making things happen.

I'd like to gratefully acknowledge cardiologist and author Dr. William Davis, who generously contributed the case study featured in Chapter 6. Thanks, Bill.

It's been a real pleasure and honor working with every member of the team at John Wiley & Sons, Canada. I feel like my book and I belong here, and I'm thankful for that.

Finally, I'd like to express my humble gratitude to the many brilliant researchers and true scientific experts whose names appear in the endnotes of this book. This book would not have been written without your efforts and I hope it does justice to your work.

1

The Calcium Paradox

I n April of 2011, nutrition researchers shocked the medical community when they published—in the prestigious *British Medical Journal* no less—the results of yet another calcium and heart health study. According to the study, women who supplement with calcium to prevent osteoporosis are at a higher risk of atherosclerosis (formation of calcium plaques in the arteries), heart attack and stroke than those who don't.[1] The outcome was clear: the increased risk of death from heart disease associated with calcium supplements outweighed any benefit to bone health. Based on this research, for every bone fracture calcium supplementation prevents, it precipitates two potentially fatal cardiovascular disease events.

This was not the first analysis to report this staggering finding; it was the *third* trial to confirm the trend.[2] Confusion quickly set in among the millions of health-conscious consumers who heard the news—and among the health care providers who recommend calcium supplements. If we don't take calcium, aren't we doomed to crumbling, fractured bones from osteoporosis? If we do take calcium, are we doomed to suffer hardening of the arteries and death from cardiovascular disease? *Is calcium killing us?* The implications were so staggering and baffling that the studies were largely swept under the rug, so don't be surprised if you didn't hear about it at first.

Exactly how did the studies' authors arrive at their unsettling conclusion, and what does it really mean? According to this research, if 1,000 women take calcium supplements for five years, at the end of that period there will be three fewer bone fractures in that group compared to a similar group of women who didn't take calcium supplements. Three fewer fractures doesn't sound very impressive and other studies do report greater fracture-prevention power. However, even if you take these results at face value and multiply those three saved fractures over the millions of women taking calcium supplements, it adds up to a meaningful benefit. Furthermore, it's the best nonprescription bet we've got. As long as there's no overwhelming drawback to popping calcium pills, then everyone should do it. And there's the rub.

The study goes on to say that in the same group of calcium takers over those same five years, there will be six more cardiovascular disease events (heart attack or stroke) than in the nonsupplement group. Six events in 1,000 women also might not sound like major cause for alarm, but it adds up too. More importantly, it's twice the number of fractures prevented, with potentially graver consequences. Curiously, the occurrence of heart attacks is not found to be dose-dependent. In other words, we do not see a greater number of heart attacks in women who take higher doses of calcium, a finding we'll explore later in this chapter. The researchers also note that the ill effect on heart health is not seen with dietary calcium from food. Given these results, the study authors make the staggering declaration that women should abandon calcium supplements.

But what about our bones? Critics of the calcium and heart health studies complain that the research generates more questions than answers. Sometimes groundbreaking research does that, but a more salient criticism is that the studies are raising questions that can't be answered by the research being done. Should we be taking calcium supplements or avoiding them like pleated pants? It turns out that the question "Are calcium supplements safe?" is the wrong one to ask. Studies looking only at calcium will never adequately answer that question—or they will inaccurately conclude that calcium is harmful. *The real question to ask is, "How can the body guide calcium safely into the bones where it helps us, and keep it away from soft tissues like arteries where it harms us?"* The answer is a long misunderstood fat-soluble vitamin called K_2.

Why You Need to Read This Book, Whether or Not You Take Calcium Supplements

The calcium question—and its surprising answer—is not just for calcium supplement takers, and definitely not just for women. Even if you rely only on food for your calcium intake, heart disease, caused by a deadly accumulation of calcium in arteries, is the number one killer of both women and men in North America. Meanwhile, osteoporosis is a major cause of disability and death in the elderly of both sexes, and calcium and vitamin D supplementation haven't helped prevent it nearly as much as we'd like. This, in a nutshell, is the Calcium Paradox: a mysterious, concurrent calcium deficiency (in the skeleton) and calcium excess (in the arteries) that underlies two major health concerns of our time, osteoporosis and heart disease. Vitamin K_2 is the key to putting calcium back in its place to remedy this calcium conundrum. This book will tell you exactly how to do it.

In the face of evidence that calcium is dangerously collecting in blood vessels while our bones are starving for the mineral that is so close by, the advice to "just stop taking calcium supplements" misses the big picture. Calcium belongs in our bones just as gasoline (environmental objections aside) belongs in the tanks of our cars. You wouldn't go to the gas station and just spray fuel all over your vehicle; you use a nozzle to put the gas where it will do the most good. Vitamin K_2 funnels calcium into bones to strengthen mineral density and fight fractures while it prevents and even removes dangerous arterial calcification. Along the way it has beneficial effects for almost every major health concern of our time, including diabetes, cancer, Alzheimer's disease, infertility, tooth decay and growing healthy children. Taking the paradox out of calcium is just the beginning of what vitamin K_2 can do for you.

There's more than a coincidental association between brittle bones and hardened arteries. Although multiple factors contribute to each disease, one common underlying mechanism unites both conditions: inappropriate calcium metabolism caused by vitamin K_2 deficiency. The very problem you are trying to stave off by taking calcium supplements predisposes you to having that calcium land in your arteries. In order to fully grasp what goes wrong in the Calcium Paradox—and how to put calcium back where it belongs—let's take a closer look at both osteoporosis and heart disease. Understanding the link between these conditions provides a framework for appreciating their connection to many other common ailments.

Osteoporosis: Calcium Deficiency

Osteoporosis is a loss of bone mineral density and thinning of bone tissue that causes bones to become more porous and prone to fracture. It is the most common bone disease, with one in five women over the age of 50 having the diagnosis, and many more having the disease that is yet to be diagnosed. Half of women over the age of 50 will experience a fracture due to diminished bone density. One in eight men over the age of 70 will develop osteoporosis, associated with a drop in testosterone that occurs around that age. There are no symptoms in the early stages of the disease; often the first sign of osteoporosis is a bone fracture due to little or no trauma.

Over time, osteoporosis can rob you of your posture—leading to a loss of height of as much as six inches—and result in a stooped stature known as kyphosis (a humped back or "dowager's hump"). Osteoporosis-related fractures, especially of the hip and spine, are a leading cause of disability among seniors. For many people, recovery from hip fracture is a lengthy and painful process. For others, a hip fracture marks the beginning of a long, complicated decline in health from which there is no recovery. Up to one-third of hip fracture patients die within one year of their fracture. Up to 75 percent of those who were independent before their fracture neither walk independently nor achieve their previous level of independent living after breaking a hip.[3]

Osteoporosis drugs are at best a dubious solution to this widespread problem. One commonly prescribed medication, alendronate sodium, has been linked to necrosis (rotting) of the jawbone, even when taken for only short periods.[4] In other individuals, bisphosphonates, the most popular class of prescription drugs developed to treat osteoporosis, have the unfortunate side effect of *increasing* the risk of bone fracture.[5] Although debate continues about the long-term safety of these medications and whether periodic drug holidays will help protect patients from serious complications, it is certain that osteoporosis isn't caused by a deficiency of prescription drugs.

What does cause osteoporosis? Although we think of our bones as solid and unchanging, the skeleton is as dynamic as any other body tissue. The amount of bone mass in the skeleton increases until the age of 30, though 90 percent of your lifetime maximum bone density, called peak bone mass, is achieved by age 20. Between age 30 and menopause, women typically experience little change in total bone mass, but not because bone tissue is static. Old bone is continuously removed and replaced by new bone to maintain a strong, healthy frame. In the first few years after menopause, women often experience a rapid loss of bone tissue due to a withdrawal of estrogen, with its bone-protective effects. The rate at which bone is lost eventually slows, but if too much bone mineral density is lost, osteoporosis results. Osteoporosis occurs when there is an imbalance between new bone formation and old bone resorption. The body may fail to form enough new bone, or too much old bone may be resorbed or both.

There are certainly genetic factors that contribute to weak, fragile bones, such as ethnicity and body type. Asians and Caucasians have it worse, and slim, small-framed women lose more bone density than their sturdier, more heavyset sisters. Exercise helps prevent osteoporosis, so women who are physically active have stronger bones. Of the factors we can control, other than being physically active, *osteoporosis is ultimately a product of how much peak bone mass you can accumulate by age 20 and how much of it you can keep after menopause.* Both of those factors are governed by vitamin K₂. The fact that osteoporosis becomes more common after menopause is not just inevitable bad luck for women. Declining estrogen levels negatively impact bone density in three distinct ways. Vitamin K₂ counteracts each of those pathological mechanisms. K₂ even affects estrogen metabolism itself.

Conventional wisdom has dictated that, since osteoporosis is characterized by a lack of calcium in the bones, adequate calcium intake is the most important remedy for the problem. Suggested calcium doses, above and beyond dietary intake, have climbed to over 1,500 milligrams daily, to only modest avail. Confusion persists and marketing hype abounds about the best, most absorbable forms of calcium that will really penetrate skeletal tissue, as if the ability to do so is a property of the calcium itself. This has led to much debate about the benefits of calcium carbonate versus calcium citrate, and created a market for calcium supplements from ridiculously exotic sources like coral reefs and desert mineral deposits.

Why don't calcium supplements cure osteoporosis? Why does this disease seem to be so stubborn? The fact is, you could be consuming and absorbing plenty of calcium from food or supplements, it's just not getting to where you need it. Even worse, it might be landing in the last place you want it: gathering in your arteries, contributing to North America's leading killer, heart disease.

Atherosclerosis: Calcium Excess

The terms "cardiovascular disease" and "heart disease" encompass many pathological conditions. They can refer to a disease of heart valves or heart muscle, or other systemic disorders that affect the heart and/or blood vessels. Throughout this book, I use the term "heart disease" to mean only coronary heart disease (CHD), also known as coronary artery disease. CHD refers to a narrowing of the blood vessels that supply blood and oxygen to the heart. This narrowing is caused by atherosclerosis, a buildup of calcium-laden plaque that slowly clogs one or more of the coronary arteries, or any artery in the body.

Atherosclerotic heart disease is explored in Chapter 4 in detail. In short, as the coronary arteries narrow, blood flow to the heart can slow down or stop. This might cause chest pain (angina), shortness of breath and other symptoms, usually when you are active. More commonly, however, the gradual narrowing of arteries goes unnoticed for years until the sudden onset of a heart attack. In Canada, myocardial infarction—a heart attack—occurs once every seven minutes, and 30 percent of all deaths in both men and women are due to myocardial infarction. Even with cholesterol screening, electrocardiograms and stress tests, the majority of heart disease cases go undetected until heart attack strikes, and 50 percent of first heart attacks are fatal.

Fighting heart disease has been a major public health concern for decades. The war against heart disease has largely dictated expert dietary advice over the last 50 years. Based on the principle that our diet—dietary saturated fat, in particular—predisposes us to heart disease, well-meaning diet dictocrats took to modifying our meals in specific ways to prevent heart disease. It wasn't particularly successful. We looked to cultures that have low rates of heart disease—French, Italian, Greek—and found them eating lots of saturated fat. We declared that a paradox and inferred that some secret ingredient, olive oil or red wine, is protecting them from the butter and egg yolks that must be killing us.

Although the lipid hypothesis, the notion that saturated fat and cholesterol cause heart disease, has been largely debunked in scientific literature,[6] it remains entrenched in popular nutrition dogma. Depending on your current awareness of the causes of heart disease, you will be pleasantly surprised or completely horrified by the list of foods high in heart-healthy K_2, discussed in Chapter 3. For now, let's just say that the French Paradox—the supposed contradiction between a rich, fatty diet and low heart disease rate—isn't such a paradox after all. And it probably isn't the red wine that's protecting those French (and Italian, Greek and Portuguese) arteries. Sure, there is some evidence to suggest that resveratrol, a compound in red grape skins, has heart health benefits, but it's nothing close to that of K_2. The shocking truth is that many of those rich, fatty "sin" foods are abundant in K_2, the only vitamin known to prevent and reverse atherosclerosis.

In 2004, the highly reputable *Journal of Nutrition* published the results of the Rotterdam Study. This population-based study, conducted in the Netherlands, evaluated almost 8,000 men and women over age 55 on their health, use of medication, medical history, lifestyle and risk indicators for chronic disease and diet. The study revealed that a high intake of vitamin K_2 from dietary sources significantly reduced the incidence of arterial calcification and risk of death from cardiovascular heart disease by 50 percent as compared to people with low dietary vitamin K_2 intake. K_2 intake was also inversely related to severe arterial calcification and so-called all-cause mortality, or death from any cause.[7] According to this study, individuals with the highest dietary K_2 will live, on average, seven years longer than their K_2-deficient counterparts.

Why Vitamin D Won't Save Us from the Calcium Paradox

Vitamin D, another fat-soluble nutrient famous for bone health, has made major headlines in the last decade. Since vitamin D is beneficial for so many diseases, doesn't taking it help protect against the Calcium Paradox somehow? Unfortunately, it does not. Calcium supplementation increases the occurrence of heart attack and stroke with or without vitamin D, showing that the latter has no protective effect here. Even worse, it's possible that the soaring popularity of vitamin D might actually be compounding the problem. Under certain circumstances, vitamin D *increases* arterial calcification. Vitamin D specifically accelerates the accumulation of arterial calcification in vitamin K_2–deficient conditions.[8] With all the good news about vitamin D, how could this be?

The news about vitamin D hasn't been all good, just the widely publicized news. We know vitamin D is beneficial for bone health. When it comes to heart health, the research has been decidedly mixed. The results are so confusing and conflicting that researchers are only just now making sense of it. Many studies indicate that vitamin D deficiency is associated with heart disease, and as vitamin D levels go up, arterial calcification decreases. Other studies show just the opposite—that higher blood levels of vitamin D are associated with more arterial plaque.[9] This double-edged sword can be partially explained by understanding what vitamin D does and doesn't do with calcium.

Vitamin D increases the absorption of calcium from the intestines, which is a good thing for bone health. Certainly, vitamin D and calcium supplementation together have been shown to increase bone density better than either one alone. However, once calcium is absorbed into the blood stream, vitamin D has no power over what happens to it, which is a potentially bad thing for heart health. Some calcium will find its way into your bones, but more of it might wind up in your arteries. Vitamin K_2 tips the balance in favor of bone *and* artery health by putting calcium in its place.

The fact that vitamin D governs calcium absorption only, then lets calcium run wild once we absorb it, explains only why excess vitamin D is bad for heart health. It doesn't account for the research showing that vitamin D deficiency is also associated with atherosclerosis, or why increasing vitamin D levels lower calcium plaque in some people. The answer to that lies in the fact that we need vitamin D to benefit from vitamin K_2 and vice versa. When vitamin D is lacking, vitamin K_2 can't do its job escorting calcium away from arteries and into bones. Chapter 7 fills in the details of this fascinating fat-soluble friendship.

This book does not dethrone vitamin D. In fact, it adds to the growing list of vitamin D benefits by placing vitamin D squarely in the "heart-healthy" category—as long as you take it with vitamin K_2. The sunshine vitamin truly is a wonder nutrient, if it has all the necessary allies needed to fulfill its potential. More vitamin D is better for heart health to a certain point, after which more is worse. Exactly where that point is depends on vitamin K_2. Having plenty of K_2 enables us to profit from vitamin D like never before. If for years you have been following expert advice by dutifully gobbling up calcium and vitamin D, vitamin K_2 not only will allow you to finally reap all the benefits of those nutrients, it might just save your life.

How Vitamin K_2 Comes to the Rescue

Vitamin K_2 works by activating a number of special proteins that move calcium around the body. Specifically, K_2 activates a protein called osteocalcin, which attracts calcium into bones and teeth, where calcium is needed. K_2 activates another protein called matrix gla protein (MGP), which sweeps calcium out of soft tissues like arteries and veins, where the mineral is unwanted and harmful. When K_2 is lacking, the proteins that depend on K_2 remain inactive. The Calcium Paradox then gradually rears its ugly head with an insidious decline in bone mineral density and an even more treacherous hardening of the arteries. When K_2 is plentiful, bones remain strong and arteries remain clear.

Throughout this book I often refer to the benefits of vitamin K_2—or the problems associated with a lack of it—by referring to the actions of the K_2-dependent proteins, especially osteocalcin and MGP. When these proteins are switched on by vitamin K_2, they actively usher calcium to and from appropriate areas of your body. When K_2 levels are inadequate, those proteins are useless and calcium wanders aimlessly, eventually taking the path of least resistance, embedding in soft tissues rather than trying to force its way into hard bone. Although the discussion may seem technical, taking a few minutes here to grasp the nature of these K_2-dependent proteins will allow you to really appreciate the amazing power and profound health benefits of vitamin K_2 that you'll learn about later in the book.

The term "protein" usually conjures up images of beef, chicken and eggs. These foods are high in essential dietary protein, but the word "protein" in the term "K₂-dependent protein" refers to something slightly different. Biological proteins are microscopic components made up of amino acids. Most biochemical reactions in any living organism occur due to the action of some protein, usually an enzyme. An enzyme is a catalyst, a protein that facilitates biological reactions.

Biological proteins need helper molecules, called cofactors, in order to work. Vitamins and minerals are cofactors. Indeed, the purpose of most vitamins and minerals in our diet is to act as cofactors for our body's proteins. Vitamin K₂ is the cofactor for an enzyme called vitamin K–dependent carboxylase. This enzyme, once and only once it is activated by vitamin K₂, alters the structure of osteocalcin and MGP to allow those proteins to bind calcium (see the sidebar "Gamma-carboxylation" for the keener-level details). Once these proteins have the ability to bind calcium, they can work wonders.

Gamma-carboxylation

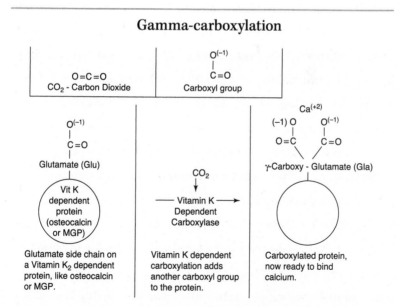

Glutamate side chain on a Vitamin K₂ dependent protein, like osteocalcin or MGP.

Vitamin K dependent carboxylation adds another carboxyl group to the protein.

Carboxylated protein, now ready to bind calcium.

The K family of vitamins activates enzymes that modify certain proteins to allow them to bind calcium. Vitamin K_1–dependent proteins are involved in blood clotting. Vitamin K_2–dependent proteins move calcium into bones and out of soft tissues like arteries, veins and skin. The process by which K_1 and K_2 activates proteins is called gamma-carboxylation. When vitamin K_2 is deficient, we say the K_2-dependent proteins are "under-carboxylated." This term is synonymous with vitamin K_2 deficiency.

Osteocalcin (also known as bone gla protein or BGP) is a biological protein found in bones and teeth. It is the most abundant protein in bone after the collagen that forms the matrix that holds calcium. Together, vitamins A and D cause special bone-building cells (osteoblasts) in our skeleton to secrete osteocalcin and use the protein to draw calcium into bone tissues. Osteocalcin isn't ready for work as soon as it is made, though. It won't work at all until its structure assumes a shape that allows it to bind calcium—and that doesn't happen by accident. It takes vitamin K_2 to activate (carboxylate) osteocalcin so it can bind the precious mineral we need to build our bones and teeth. Inactive (under-carboxylated) osteocalcin is powerless; it won't bind calcium and it won't build bone tissue. In fact, measuring your level of inactive osteocalcin is a handy way to assess K_2 deficiency; if K_2 is lacking, more useless osteocalcin will be hanging around. Vitamin K_2 deficiency testing is discussed in Chapter 6.

In addition to building bone density, osteocalcin, produced by our bones and teeth, plays some unexpected roles in health. For example, new research shows that osteocalcin acts as a hormone that causes the pancreas to secrete more insulin and increases sensitivity to insulin at the cellular level.[10] Insensitivity to insulin (also called insulin resistance) is at the heart of the epidemic of obesity and type 2 diabetes that now plagues the Western world. This new understanding of osteocalcin confirms that our skeleton is not just inert scaffolding. It is an endocrine gland that plays a role in the prevention of diabetes. More to the point, *vitamin K₂, essential for osteocalcin to function, is likely a critical nutrient in preventing and treating type 2 diabetes.* In addition to osteoporosis and coronary heart disease, a decline in dietary K₂ has no doubt contributed significantly to the obesity and diabetes crisis that is upon us.

Another newly identified and even more astonishing role for osteocalcin is in male fertility. Men's bones, via secretion of osteocalcin, actually help regulate testosterone production.[11] This impacts sperm production and survival in the testes. No doubt this is the mechanism that underlies the traditional wisdom of many cultures that men preparing to become parents consume plenty of K₂-rich foods. Indeed, K₂ plays so many roles in male and female fertility, perinatal wellness and growing, healthy children that I devote much of Chapter 5 to that information.

Just as K₂-activated osteocalcin directs calcium into bones and teeth, where it is helpful, osteocalcin's counterpart, matrix gla protein (MGP), escorts calcium out of the areas where it is harmful, like arteries and veins. MGP resides in numerous body tissues, including bones and that of the heart, kidneys and lungs. As with osteocalcin, vitamin D stimulates MGP production. Mice that lack MGP altogether die within two months of birth as a result of massive arterial calcification that leads to blood vessel rupture.[12] In animals and humans, when MGP is present but remains mostly or even partially inactive due to lack of K₂, the same calcification process occurs, just milder and more slowly.

Vitamin K_2–activated MGP is the strongest inhibitor of tissue calcification presently known. Its pivotal importance for cardiovascular health is demonstrated by the fact that there seems to be no effective alternative mechanism for preventing calcification in blood vessels.[13] In other words, when vitamin K_2 is deficient, the calcium plaque buildup of atherosclerosis is unavoidable—and this is where things get spooky.

When I mentioned earlier that unsupervised calcium eventually becomes embedded in arterial tissue, it may have sounded like a passive process. That's a popular misconception. If inappropriate calcification happened by chance, we'd expect the calcium and heart health studies to show more cardiovascular disease events in people who take higher doses of calcium, but they don't.[14] Furthermore, if calcification were an indiscriminate process, we'd also probably see randomly calcified bits here and there throughout the body, but we don't. The arteries aren't the only tissues that can develop ectopic (out of place) calcification, but they are usually the first and most sensitive areas.

For a long time, the prevailing notion of the arterial calcification process was that it was a passive affair associated with advanced atherosclerosis. In other words, when fatty material had clogged arteries for long enough, it would eventually harden, due to calcium deposits, because there was no particular mechanism to prevent calcium from building up. Now we know that isn't the case. Calcium is actually present in a fairly consistent ratio—occupying about 20 percent of the volume of an arterial plaque—from the very early stages of the plaque formation. That is why greater calcium intake doesn't translate into more heart disease, as shown by the calcium and heart health studies that sparked this discussion.

Calcium doesn't drift into arteries by fluke. Minerals deposit into atherosclerotic plaques by an active process that mirrors bone formation.

Artery wall calcification due to atherosclerosis frequently contains fully formed bone tissue, including marrow.[15] Osteoblast- and osteoclast-like cells within the artery go haywire and actually form tissue that, under a microscope, is indistinguishable from bone. Arterial calcification is really a process of ossification—bone building.

In a sense, this weirdly inappropriate bone-building phenomenon has protected us from even more serious effects of widespread calcium supplementation. Rather than having all the unusable calcium clogging our arteries, only a portion of it winds up there. That's good news, at least, but the fact remains that we need calcium to build our bones, and we can't afford to sacrifice our heart health to get it. What causes dormant, bone-building-type cells in our blood vessels to malfunction and create bone tissue where it should not be? Vitamin K_2 deficiency. By activating MGP, vitamin K_2 ensures calcium contributes to bone tissue buildup only where it should and prevents bone from being laid down where it shouldn't be.

K_2-activated MGP doesn't just prevent atherosclerosis, it reverses life-threatening arterial plaque. Yes, you read that right. It is actually possible to lessen plaque burden by stimulating more of your MGP to actively sweep calcium away. Animal studies show a 37 percent decrease in arterial calcium content after only six weeks on a vitamin K_2–rich diet. This benefit is mediated entirely by K_2-activated MGP.[16] MGP is now being used as a biochemical marker for arterial calcification. Blood tests that measure your level of active versus inactive MGP can accurately predict how much calcium plaque you have. Vitamin K_2 supplementation increases active MGP levels in humans in a dose-dependent manner: more K_2 means more K_2-activated MGP.[17] This, in turn, means less arterial calcification.

Not worried about heart disease because your cholesterol isn't high? Keep in mind that heart disease is not called a silent killer for nothing. Ninety percent of cases go undetected until heart attack strikes.[18] Whether your cholesterol is high or low, what really matters is whether calcium plaque is building up in your arteries, leading to a potentially fatal blockage. Since heart disease is the number one cause of death in North America— and just focusing on cholesterol will lead you astray in cardiovascular disease prevention—it's worth learning how to get K_2 back into your diet.

Unlike osteocalcin, which, with a few notable exceptions, is mostly confined to bone tissue, MGP pops up throughout the body. It is found in bones, blood vessels, the heart, lungs, kidneys and cartilage. Uncarboxylated, K_2-deficient MGP is associated with disease in each of those areas. Curiously, many types of malignant tumors also produce MGP for reasons unknown. Probably not coincidentally, K_2 deficiency fosters cancer growth. Scientists continue to explore the role of MGP and vitamin K_2 in multiple health conditions, and amazing benefits are still emerging.

The Calcium Cycle of Life

There is a fascinating interplay between fat-soluble vitamins, calcium metabolism and the seasons, which conveys the interconnectedness of osteoporosis and atherosclerosis. Both arterial calcifications and bone density vary according to an annual cycle. Arterial plaque builds up in the wintertime and diminishes slightly in the summer, a phenomenon explored further in Chapter 4. Bones do the opposite. Bone mineral density loss occurs almost exclusively during the winter, with virtually no loss in the summer.[19] Unfortunately, the lost bone mineral content isn't usually regained in summer, but bone density at least remains constant at that time. On an annual basis, then, calcium is lost from the skeleton at the same time it is accumulating in arteries. Supplementing with calcium and vitamin D prevents winter bone loss, but popping calcium pills during prime plaque-building season is risky. Ah, the Calcium Paradox thumbs its nose at us once again.

The complete answer to this cyclical calcium riddle will unfold throughout the book. It has to do with humans' delicate interconnectedness to the sun and the earth. For the moment, suffice it to say that vitamin K_2 cooperates with other fat-soluble nutrients so that we may benefit from calcium without risking harmful side effects. There has always been an annual variation in osteoporosis and atherosclerosis—this is the ebb and flow of life. Understanding this pattern and what causes it provides a framework for understanding what a healthy diet really is and when vitamin supplements are needed. Supplementation on top of excellent nutrition may help us cheat death just a little bit, or at the very least buck the seasonal trend.

We All Need More: Vitamin K_2 Deficiency Is Widespread

Now that you know how vitamin K_2 works in the body, what's to say you don't already have enough of it? If you know you have either osteoporosis or heart disease (or both), K_2 deficiency is a given—but keep in mind that most people are unaware they have those conditions until disaster strikes. If you are menopausal or have a history of cancer, infertility, varicose veins or diabetes, the likelihood of K_2 deficiency is very high, since those conditions are all associated with an increased requirement for or deficiency of the nutrient. For other hints that you might be lacking K_2, scan the list of K_2-deficiency conditions below.

Conditions associated with vitamin K_2 deficiency:

- osteoporosis
- atherosclerosis
- increased risk of cancer (including breast, prostate, liver)
- diabetes
- varicose veins
- wrinkles
- dental cavities
- Crohn's disease
- kidney disease
- narrow, crowded dental arch
- adolescence

Even without any of these health concerns, another very compelling factor points to the strong probability that you are K_2 deficient: according to recent research, most people are.[20] A 2007 study revealed that the majority of "apparently healthy" individuals have substantial levels of under-carboxylated osteocalcin and matrix gla protein (MGP), caused by vitamin K_2 deficiency.[21] In other words, most people do not have adequate vitamin K_2 levels to fully activate the proteins needed for optimal bone and heart health. If you can be K_2 deficient and apparently healthy, then what's the big deal? Based on the most current understanding of how and why we grow old, the triage theory of aging, undetected vitamin K_2 deficiency now will take its toll later in life. Poor vitamin K_2 status must be regarded as a serious risk factor for increased postmenopausal bone loss, artery calcification, diabetes, end-stage kidney disease and aging itself.

K_2 has a better-known sibling called K_1, whom we'll meet in Chapter 2. The main role of K_1 is in blood coagulation, not calcium metabolism.

In healthy people, 100 percent of K_1-dependent proteins are activated by vitamin K_1. In contrast, a varying percentage of osteocalcin and MGP is left inactivated by K_2 in the same people. While most everyone gets the vitamin K_1 they need for proper blood clotting, researchers rarely find an individual with enough vitamin K_2 to meet their calcium-metabolism needs. As important as coagulation is, vitamin K_1 has no effect on heart disease risk, judging by the Rotterdam Study, and little effect on bone strength. Almost everyone is lacking vitamin K_2; we differ only in the degree of deficiency. It wasn't always this way. Exactly how we got into this sorry state—and how to get out of it—is explained in Chapter 3.

Critics of the calcium/heart health studies point out, and rightly so, that saying there is an increase in heart attacks and strokes in women who take calcium is not the same as saying calcium supplements cause heart attacks and strokes. That is true, but it's kind of like arguing that bullets aren't harmful. Calcium supplements are the ammunition in the weapon of vitamin K_2 deficiency. Should you discontinue calcium supplementation altogether to avoid heart disease? Not necessarily. With sufficient K_2, however, you might benefit enough from the calcium naturally present in a healthy diet to not need calcium supplements.

For those with osteoporosis, calcium might still be needed. What about vitamin D? As I explain in Chapter 7, taking vitamin D increases the body's need for K_2. By jumping on the vitamin D megadose bandwagon, you are compounding the potential danger of calcium supplementation if you are not also taking K_2. Conversely, vitamin D boosts the requirements and potential benefits of K_2. You can profit from vitamin D without increasing your risk of inappropriate calcification by having a balanced intake of all the fat-soluble vitamins, including K_2.

The discovery of vitamin K_2 is the final piece in the nutritional puzzle of many widespread diseases. How is it possible that we overlooked this incredibly important vitamin until now? It was, in part, a case of mistaken identity. The next chapter tells the story of how K_2's more popular sibling distracted our attention for decades, and how fascinating research about K_2 was hidden in plain sight for over seven decades. Read on to learn the answer to a 70-year-old mystery, and how you can make sense of conflicting information to ensure your diet supplies the kind of vitamin K that counts.

2

The Undiscovery and Rediscovery of Vitamin K$_2$

E ven though most of the world is just hearing about vitamin K_2, it isn't new. Scientists discovered K_2 70 years ago; they just didn't know what it was, or—more accurately—they thought it was something else. Misconceptions about this vital nutrient persisted for decades, and we failed to recognize its unique actions, food sources and deficiency symptoms. Confusion about the nature of K_2 persists to this day, in large part because of the lingering effects of its botched discovery. This chapter explores that story, and reveals that vitamin K_2 is the answer to a 70-year-old mystery. I'll also set the record straight about the difference between K_2 and its sister molecule, vitamin K_1. But first, in order to clarify what K_2 is—and what it isn't—it's helpful to understand how we came to know about it at all.

A Brief History of Vitamin K: A Tale of Two Nutrients

Vitamin K was discovered in the early 1930s by Danish biochemist Henrik Dam (1895–1976). Dam was studying another fat-soluble nutrient, cholesterol, and working with laboratory chickens on very-low-fat diets. Mysteriously, some of the chicks in the study became ill, developing severe internal hemorrhages because their blood was unable to clot as usual. Dam found that the problem could be prevented by giving the chicks specific foods, particularly greens and liver, yet the clotting problem did not match up to any known nutrient deficiency.

Eventually, the factor required for clotting was identified and named vitamin K because, in Dam's own words, "the letter K was the first one in the alphabet which had not been used to designate other vitamins, and it also happened to be the first letter in the word 'koagulation' according to the Scandinavian and German spelling."[1] Almost a decade later,

American researcher Edward Doisy (1893–1986) succeeded in isolating vitamin K and thereby positively identified the nutrient and its structure. In 1943, Dam and Doisy shared the Nobel Prize in physiology and medicine for the discovery of the "coagulation nutrient," vitamin K_1. And this is where things went sideways for vitamin K_2.

Both Dam and Doisy, as well as other researchers around the world, recognized that vitamin K appeared in two distinct forms, designated K_1 and K_2. However, although both forms were discovered and characterized over the course of the 1930s, three fundamental misunderstandings about these nutrients persisted for the next 70 years. First, K_1 and K_2 were considered to simply be structural variations of the same vitamin and not unique nutrients with discrete properties. Second, blood clotting was thought to be their only role in the body. Third, vitamin K deficiency was assumed uncommon and obvious, since it would manifest as some kind of bleeding disorder. These last two assumptions are accurate when it comes to K_1, but highly inaccurate when it comes to K_2.

Although it was not pursued, there must have been at least a notion among scientists studying vitamin K that the nutrient somehow had health impacts beyond coagulation. In his 1946 Nobel lecture, Henrik Dam made a passing reference to early inklings that vitamin K might play a role other than blood clotting, but then dismissed the idea: "It . . . seems unlikely that vitamin K, as such, should play any role in the prevention of caries."[2] If by "as such" Dam meant vitamin K_1, then he was right: phylloquinone (K_1) does not play any direct role in preventing dental cavities. But K_2, menaquinone, plays a big one.

Incredibly, the discovery of the first vitamin K–dependent activity unrelated to blood clotting didn't occur for almost another 30 years—and it was a major milestone in changing the fundamental perception of vitamin K. In 1975, researchers at the Harvard Medical School discovered the vitamin K_2–dependent protein osteocalcin, which we now know to be a critical factor in drawing calcium into bones and teeth to prevent osteoporosis and dental cavities.[3]

Despite this radical discovery, it would be yet another 20 years until the scientific community realized that vitamin K is "not just for clotting anymore."[4] In 1997, researchers reported that the nutrient was required for two critical physiologic processes unrelated to coagulation: ensuring healthy calcium deposition in bones and preventing calcification of arteries that leads to premature death. The implications of this finding were astounding. For the first time, scientists had identified a single nutritional compound that governed the appropriate deposition of calcium in the body. The puzzle of two widespread but seemingly unrelated diseases, osteoporosis and atherosclerosis, was being solved. So why didn't you hear about this 15 years ago?

Although K_2's role in preventing these major diseases is now obvious, in the 1990s we still didn't quite get the relevance of this nutrient. After all, even though K_2 was clearly necessary for optimal bone and heart health, little evidence existed to suggest that a lack of this nutrient was a common problem. The most surprising revelation of all was finally made in 2007: vitamin K_2 deficiency is, in fact, very widespread, and this is having a major impact on human health.[5] Scientists are still grappling with the full ramifications of this plight. We know that osteoporosis, atherosclerosis, cancer and other serious health conditions are implicated. Research about the amazing benefits of K_2 is still pouring in.

The Mysterious Activator X

There's a little more to the history of vitamin K_2 than its bungled and delayed discovery by the mainstream scientific community. An astounding body of evidence that illuminates our modern understanding of menaquinone was actually published in 1939, four years before Dam and Doisy accepted their Nobel Prize. For decades, this wealth of knowledge sat right under the collective nose of scientists and nutrition experts, undiscovered because its author, who didn't know the identity of the vitamin he was studying, simply referred to the nutrient as "X." Furthermore, the author's formal training made him an unexpected source of groundbreaking nutritional research. He was, after all, a dentist.

Dr. Weston A. Price was not your average dentist. He has been called the "Charles Darwin of nutrition" thanks to his discoveries about the causes of dental cavities and chronic disease. Dr. Price's work, which took him around the world in search of the origins of illness, resulted in the discovery of a new fat-soluble nutrient that he named "activator X." Price demonstrated that the nutrient clearly played a critical role in health and a lack of it would produce illness in a very predictable pattern. For decades, the identity of activator X remained a mystery and the subject of debate in the realms of medicine and nutrition. Now we know it is vitamin K_2. The fascinating life and work of Dr. Weston Price provide an abundance of original, evidence-based information about the actions and health benefits of vitamin K_2 to which modern research is just catching up. Understanding his findings provides a framework for appreciating the full spectrum of remarkable healing properties of vitamin K_2.

Born in 1870 near the village of Newburgh, Ontario, Weston Andrew Price moved to Ohio in the 1890s, settling in Cleveland, where he practiced dentistry for the next 50 years. But, right from the beginning, something bothered Dr. Price about his practice: it was too busy. It didn't seem right to him that so many people had such bad teeth. Price reckoned that this wasn't natural. He suspected that something about people's modern, industrialized lifestyle was having a seriously negative impact on dental health and general well-being. And so, in 1925, after three decades of treating people whose teeth and bodies were plagued with the common maladies of the modern day, Dr. Price and his wife, Florence, embarked on a series of extensive and often hazardous expeditions to find people around the world who were truly healthy and to determine what made them so.

Using Indiana Jones–era modes of transportation, the Prices made their way to remote corners of the globe: frigid Alaska, the most primitive regions of Africa, faraway Australia and New Zealand, the idyllic archipelagos of the South Pacific, the windswept Outer Hebrides (an isolated chain of islands off the west coast of Scotland), barely accessible mountain villages of Switzerland, the deserts of the Andean Sierra and the jungles of the Peruvian Amazon. There Dr. and Mrs. Price found groups of people who, cut off from the influence of the modern world and without toothbrush or paste, were, simply put, healthy. The world over, the Prices found communities of traditional people who had no need for dentists—indeed, had little need for doctors of any kind. Instead, they displayed exceptional immunity to the serious afflictions that plagued the modern world. Dr. Price noted that they were able to maintain this vibrant health for a lifetime, "so long as they were sufficiently isolated from our modern civilization" and followed the ancestral diet that had sustained their people for generations. If, instead, individuals from the tribe lost this isolation and began to consume foods of modern civilization, things changed.

Without exception, Price found that when these previously healthy people adopted a modern diet—either because they left their isolated home to live in more urban areas or trade route developments brought the modern foods to them—they experienced a predictable and specific pattern of decline in their health. First, dental decay would set in. Where cavities had been unknown before, people would develop one, then several, and sometimes mouths full of rotting teeth. Then came the gum disease. Although today dental health is primarily considered an issue of dental hygiene, tooth and gum disease emerged in these individuals even though there had been no change in dental hygiene habits. Dental hygiene as we know it had not previously been necessary. More seriously, there is a predictive relationship here that was, it seems, better appreciated in Price's time and that is only now being rediscovered: tooth decay and gum disease are harbingers of heart disease.[6]

What was even more disturbing than the emergence of dental disease where it had not existed before was the equally predictable pattern of chronic disease seen in the offspring of those who adopted the modern diet. Where the parents had broad, beautiful faces, the first generation born after the introduction of modern foods had narrowed dental arches that housed crowded, crooked teeth. These children were also prone to a number of other now-common ailments, including increased susceptibility to infections, and even behavior issues. In many groups the process of birthing became much longer and more difficult as well. Price remarked that most cultures he studied observed special feeding practices and reserved sacred foods for both men and women approaching their childbearing years, as well as for growing children. Almost every culture also had customs or taboos around how often children should be born. They practiced the spacing of children so that mothers could replenish their nutrient stores for subsequent children. Apparently, traditional wisdom had a prescription for producing healthy kids. When this wisdom was abandoned in favor of the modernized diet, problems set in.

In the photos on page 31 you can see the typical broad, well-proportioned faces of healthy indigenous people. The strikingly beautiful teeth and square jaws in both men and women are now only seen in supermodels, some professional singers and elite athletes. The relative facial proportions are similar in healthy people around the world. Upper, middle and lower thirds of the face are approximately equal. The width of the jaw is about the same as the width of the forehead. Facial symmetry is the norm. Wherever indigenous people were sufficiently isolated from industrialized society and consuming only traditional foods, Dr. Price encountered villages full of adults and children with perfectly straight, healthy teeth and wide, attractive faces to match.

The next set of photos show the typical facial changes caused by a modernized diet. The most obvious defects are noticeable in the teeth, which are crowded because the lower third of the face is underdeveloped in these children. This is not due to heredity. The boy in the upper left photo of the traditional faces is the eldest child of an Australian Aboriginal family. He inherited his ample grin and straight teeth from his parents, who were born in the bush. At the time these pictures were taken, the family had moved to a reservation and were living on the imported foods provided by the government. You can see the effect this had on his little brother, the second child born into the family, shown in the upper left photo of the modern faces.

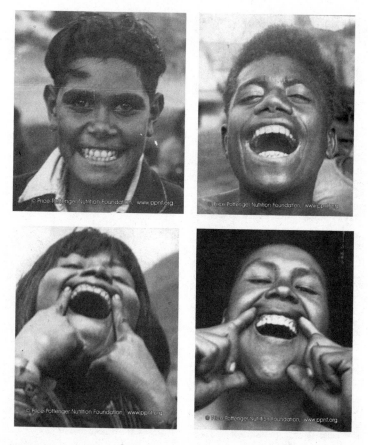

Traditional faces
Upper left, firstborn son to an Australian Aboriginal family. Upper right, typical Melanesian boy. Lower left, typical Amazon Indian. Lower right, typical Indian of the Peruvian Andes.

We see the same phenomenon in the Amazonian sisters shown in the lower left images of the two sets (traditional and modern) of photos. The firstborn has perfectly formed dental arches and straight teeth. Her little sister does not have enough room in her mouth for all of her adult teeth, so they are misaligned.

The children in the upper and lower right photos of each set respectively are not related, but the modern faces reveal common variations of

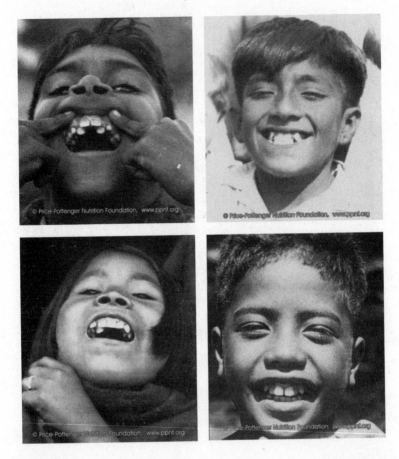

Modern faces
Upper left, second-born son to Australian Aboriginal family. Upper right, coastal Peruvian Indian boy whose parents had straight teeth. Lower left, Amazon Indian born after family adopted modern diet. Lower right, Samoan boy born after introduction of processed foods to parents' diet.

nutrient-related facial underdevelopment. Since the dental arch is smaller than necessary to fit a full set of adult teeth, the teeth come in staggered. At puberty, the incisors and canines jockey for position. The canines usually lose and are forced to erupt in front of or behind the normal arch. The canines are also pointier because they have calcified prematurely.

What were these "foods of modern commerce," as Price put it, that seemed to have such a toxic effect on those who ate them? Food that could be transported long distances without spoiling: white flour, white sugar, white rice, vegetable fats, canned goods and other processed, refined, devitalized fare. In other words, the elements that constitute the foundation of our industrialized diet. While we may recognize these are not the most nutritious choices on the menu, we rationalize that as long as they are eaten in "moderation," they can be part of a healthy diet. Yet the observations made of thousands of people from traditional cultures suggests otherwise.

But surely these modern foods must have contained some noxious substance that caused the predictable ailments? Price came to recognize that it was not the presence of some "injurious factors" that was responsible for the inevitable physical decline; rather, it was "the absence of some essential factors"[7] that was robbing people of robust good health when they habitually consumed modernized foods. To test this theory, Price performed chemical analyses on thousands of samples of traditional foods for their nutritional content and compared them with food samples from the American diet of his day. Once again, a clear pattern emerged.

Price found that the diets of healthy traditional people contained at least four times more minerals and water-soluble vitamins than the standard American diet of the 1930s. What was more surprising is that the traditional fare provided *at least 10 times more fat-soluble vitamins* than the average industrialized diet. It was to these fat-soluble nutrients that Price turned his attention.

Dr. Price recognized fat-soluble vitamins to be the foundation of health-preserving traditional diets. He called these nutrients "catalysts" and "activators" because the body needs them in order to make use of all other dietary nutrients, such as protein, minerals and water-soluble vitamins. He wrote, "It is possible to starve for the minerals that are abundant in the foods eaten because they cannot be utilized without an adequate amount of the fat-soluble activators."[8] Price was describing what we now recognize as vitamin A and D's ability to act as hormones. These nutrients work at the cellular level, stimulating our DNA to produce proteins that will employ all other nutritional compounds (as cofactors) to enhance our well-being. (For more on the profound importance of vitamins A and D, see Chapter 7.)

Aside from vitamins A and D, which were known by the 1930s, Price identified the presence of another fat-soluble "activator" in many of his test samples. Since he could not identify this new nutrient, he simply named it "activator X." This fat-soluble vitamin was distinct from the known fat-soluble vitamins, and it clearly had profound effects on the health of teeth and bones. Price found the substance in fish eggs, egg yolks and some organ meats, but especially in the butterfat of cows eating rapidly growing green grass. It was from this latter source that Price created an oil that was rich in activator X.

Dr. Price began to use the high-vitamin butter oil as the cornerstone of a nutritional protocol for treating dental cavities. Indeed, this dentist soon stopped drilling and filling teeth altogether, except in cases where extreme decay called for temporary fillings to ease pain. For most patients, in lieu of filling, he relied entirely on a dietary treatment to restore the teeth of hundreds of patients with active dental cavities and of several patients with fractured bones that were previously slow to heal. He documented his success by publishing many examples of dental X-rays before and after the nutritional treatment. The results are astonishing. Activator X obviously had a remarkable healing effect on bones and teeth.

For years, physicians and nutritionists alike pondered and debated the identity of this mysterious "X factor." One expert proposed that the fat-soluble substance was essential fatty acids. Another expert later refined that hypothesis to specify eicosapentaenoic acid (EPA), a particular type of essential fatty acid, although the properties of EPA never adequately matched the nutrient described by Price. It was only in 2007 that the mystery was finally solved. Activator X is vitamin K_2. The sidebar "Characteristics of activator X and vitamin K_2" summarizes the identifying features and similarities between what Price discovered about activator X and what we know today about vitamin K_2.

Characteristics of activator X and vitamin K_2

Activator X	Vitamin K_2
Found in the butterfat of mammalian milk, the eggs of fish and the organs and fats of animals.	Found in the butterfat of mammalian milk and the organs and fats of animals. Analyses of fish eggs are not available.
Synthesized by animal tissues, including the mammary glands, from a precursor in rapidly growing green grass.	Synthesized by animal tissues, including the mammary glands, from vitamin K_1, which is found in association with the chlorophyll of green plants in proportion to their photosynthetic activity.
The content of this vitamin in butterfat is proportional to the richness of its yellow or orange color.	Its precursor is directly associated with beta-carotene, which imparts a yellow or orange color to butterfat.
Acts synergistically with vitamins A and D.	Activates proteins that cells are signaled to produce by vitamins A and D.

Plays an important role in reproduction.	Synthesized by the reproductive organs in large amounts from vitamin K₁ and preferentially retained by these organs on a vitamin K–deficient diet. Sperm possess a K₂-dependent protein, osteocalcin.
Plays a role in infant growth.	Contributes to infant and childhood growth by preventing the premature calcification of the cartilaginous growth zones of bones.
Plays an essential role in mineral utilization and is necessary for the control of dental caries.	Activates proteins responsible for the deposition of calcium and phosphorus salts in bones and teeth and the protection of soft tissues from calcification.
Increases mineral content and decreases bacterial count of saliva.	Is found in the second-highest concentration in the salivary glands, and is present in saliva.
Intake is inversely associated with heart disease.	Protects against the calcification and inflammation of blood vessels and the accumulation of atherosclerotic plaque.
Increases learning capacity.	The brain contains one of the highest concentrations of vitamin K₂, where it is involved in the synthesis of the myelin sheath of nerve cells, which contributes to learning capacity.
Deficiency during pregnancy causes characteristic underdevelopment of the face in children, leading to crowded, crooked teeth.	Essential for proper facial development; deficiency causes identical facial underdevelopment, resulting in crowded adult teeth.

Adapted from: Masterjohn C. On the trail of the elusive X-factor: a sixty-two-year-old mystery finally solved. *Wise Traditions* 2007, volume 8, number 1, pp. 14-32.

Dr. Price deduced that the real problem of modern chronic disease was that nutrient-bereft, white foods displaced the nutrient-dense fare that nourishes the body. The processed foods provide only empty calories. When the body lacks sufficient fat-soluble nutrients to attract minerals to their proper place, demineralization of teeth and bones ensues. If he'd had the necessary technology at his disposal, Price would have seen that not only was calcium being lost from the bones and teeth but, more seriously, it was getting lodged in soft tissues like arteries. Instead, he recorded the effects of changes in vitamin K_2 intake in modern diets and showed that death from cardiovascular disease varied in an almost perfectly inverse relationship. When K_2 intake increased, cardiac mortality decreased and vice versa. The Calcium Paradox was lurking long before calcium supplements came into the picture.

Although Price's primary quest had been to find the cause of tooth decay (which, the doctor remarked, "was established quite readily as being controlled directly by nutrition"[9]), it rapidly became apparent that the physical ailments brought on by a modernized diet extended to the whole body. Since we are now all born into societies where crowded, decayed teeth are the norm, and chronic disease is accepted as "just a part of aging," it is challenging to perceive the effects of inadequate nutrition for what they are. Nutrient deficiencies have compromised human health for so long, we are no longer in a position to recognize it. From Price's unique perspective, this was self-evident.

Price's fieldwork could scarcely be repeated today. Populations untouched by modern civilization have all but disappeared. A factor that made the timing of Price's work so precarious was the availability of a key piece of technology, the camera. Price documented his observations with hundreds of remarkable photographs, which he published alongside his findings in the pivotal work *Nutrition and Physical Degeneration*. Image after image establishes the fine facial development and apparent radiant health of the so-called primitives. Similarly, the narrow jaws and deformed facial structures of those on a modern diet reveal an alarming pattern that is obvious, even to the untrained eye. More hauntingly, the specific pattern of underdeveloped facial form is one that is now very common.

Although nutrient-poor white foods crowd out healthier fare, restoring our nutrient intake is not as simple as just avoiding white flour and sugar. Likewise, the lessons of Price's work cannot be summed up with a prescription for one nutrient. The sidebar "Principles of traditional diets" briefly summarizes what he learned about traditional, health-promoting diets. And so, in a meaningful way, the discovery of vitamin K₂ was really made by a dentist from Ontario. His meticulous research establishes a body of evidence that points to the benefits of vitamin K₂ that modern science has yet to explore. Throughout this book I balance the current state of vitamin K₂ knowledge based on recent studies with relevant implications from Price's profoundly significant work. (For an absorbing, if lengthy, read on medical anthropology and nutrition science, I recommend the eighth edition of *Nutrition and Physical Degeneration*, by Weston A. Price, available at www.ppnf.org—the Price-Pottenger Nutrition Foundation.)

Principles of traditional diets

Following the example of healthy traditional diets from around the world, Price identified nutrient-dense foods available to modern North Americans. Here is a summary of his principles for a nourishing diet:

- Eliminate sugar, starch and white flour; they provide only empty calories and displace food with greater nutrient value from the diet.
- The greatest challenge is to consume foods that provide adequate fat-soluble vitamins; fish, seafood and cod liver oil are excellent sources.
- Milk and dairy products are also highly nutritious and will provide plenty of fat-soluble vitamins, providing the cow has some green grass in her diet.
- Raw vegetables are too bulky to provide nutrients efficiently; they should be used in moderation. Cooked vegetables, especially in soups, are more effective at providing concentrated vitamins and minerals. Legumes, particularly lentils, are among the most nutritious vegetables.
- Eat whole grains, always freshly ground. Much of the nutrient value of whole grains is lost by oxidation if they are not prepared and eaten within a day or two. A cooked cereal from freshly cracked oats is a good choice.

Finally Understanding Vitamin K₂ After 70 Years: What It Is and Isn't

As you are now aware, vitamin K is not a single nutrient but a family of fat-soluble vitamins. As with the water-soluble family of B vitamins, in order to make any accurate statement about vitamin K, you need to differentiate exactly which member of the family you are talking about. Only two members of the K family are useful for general health, and throughout this book I refer specifically to one or the other, except in the rare instances when my statement applies to both forms of the nutrient. The sidebar "K_3–K_7" describes a few synthetic cousins in the K clan, whose usefulness is mostly limited to professional or industrial purposes.

If vitamin K_2 is the nutrient we have long ignored and need to pay more attention to, why am I devoting any time to K_1? It's important to have a basic understanding of the difference between K_1 and K_2 so that when you read "informative" magazine articles or a product label that states, "Green vegetables are a great source of bone-building vitamin K," you'll know exactly why that is bunk. There is a small amount of overlap linking K_1 and K_2 and important distinctions to be made between them. Confuse the two, as scientists did for 70 years, and you may end up missing out.

K_3–K_7

In addition to K_1 and K_2, there are a few other members of the K family, although they are synthetic and not essential nutrients (essential nutrients are nutritional elements required for normal body functioning that cannot be synthesized by the body at all or cannot be made in amounts needed for good health and so must be obtained through our diet):

- K_3 (menadione): considered a synthetic version of vitamin K, although intestinal bacteria can produce minute amounts of it from K_1. In the United States, the Food and Drug Administration has banned its use in nutritional supplements due to liver toxicity in humans, but it is sometimes used in pet food. K$_3$ has been studied for its anticancer effects.[10]

- Vitamin K$_4$ (menadiol): administered by injection to treat hypo-prothrombinemia, a bleeding disorder caused by a deficiency of clotting factor prothrombin.

- Vitamin K$_5$ (4-amino-2-methyl-1-naphthol hydrochloride): studied as a preservative and antifungal agent for commercial purposes.

- Vitamin K$_6$ to K$_7$: vitamin K can be manipulated chemically to pro-duce many synthetic variations, some of which may one day prove to have usefulness in health care. (A note about the designation "vitamin K$_7$": some authors incorrectly refer to the bone-building benefits of "vitamin K$_7$" when what they really mean is MK-7, short for menaquinone-7, which is actually a form of vitamin K$_2$.)

Vitamin K$_1$: Greens and Clotting

The role of vitamin K$_1$, also known as phylloquinone, is to activate special proteins, called clotting factors, which allow the blood to form clots. Several clotting factors are dependent on vitamin K$_1$ for their function, and they are activated by phylloquinone in a complex system called the "clotting cascade." This life-saving mechanism prevents us from bleeding to death from, say, a paper cut. In many countries, newborns are given an injection of synthetic vitamin K$_1$ at birth to activate this system and prevent a rare bleeding disorder that can occur in early infancy.

Phylloquinone is present in all photosynthetic plants—in other words, green plants that derive energy from the sun. Chlorophyll, the pigment that imparts the green color to vegetation, contains essential phylloqui-none. K$_1$ plays a critical role in energy production within the plant cells by carrying electrons inside the cell membrane, much in the same way as coenzyme-Q10 (CoQ10) does in humans. Both phylloquinone and ubiquinone, aka CoQ10, have similar structures.

Not surprisingly, phylloquinone is abundant in green leafy vegetables (the prefix "phyll" comes from the Greek word for leaf). Excellent dietary sources of K_1 include kale, collards, spinach, turnip greens, beet greens, broccoli and brussels sprouts. Most fruits, vegetables and nuts contain some phylloquinone. The recommended daily intake is a mere 90 micrograms per day for women, 120 micrograms for men. The table below lists the vitamin K_1 content of various vegetables and fruit.

Vitamin K_1 content of selected foods

Food	micrograms
Kale, frozen, boiled and unsalted, 1 cup (4 1/2 oz)*	1,146.6
Collards, frozen, chopped, boiled and unsalted, 1 cup (7 oz)	1,059.4
Spinach, frozen, chopped or leaf, boiled and unsalted, 1 cup (7 oz)	1,027.3
Broccoli, boiled and unsalted, 1 cup (5 1/2 oz)	220.1
Brussels sprouts, boiled and unsalted, 1 cup (5 1/2 oz)	218.9
Parsley, fresh, 10 sprigs (1/3 oz)	164.0
Noodles, enriched, egg or spinach, cooked, 1 cup (5 1/2 oz)	161.8
Green leaf lettuce, 1 cup (2 oz)	97.2
Broccoli, cooked, 1 spear (1 oz)	52.2
Spinach, raw, 1 leaf (1/3 oz)	48.3
Blueberries, frozen, 1 cup (9 oz)	40.7
Celery, raw, 1 cup (4 oz)	35.2
Broccoli, raw, 1 spear (1 oz)	31.5
Kiwifruit, 1 medium (2 1/2 oz)	30.6
Avocado (1 oz)	6.0

Source: Adapted from USDA National Nutrient Database for Standard Reference, Release 17. Vitamin K (phylloquinone) measure (μg) content of selected common foods, sorted by nutrient content. www.nal.usda.gov/fnic/foodcomp/Data/SR17/wtrank/sr17w430.pdf

* Measures have been rounded to nearest half ounce.

Vitamin K₁ Deficiency Is Rare and Obvious

Don't worry about adding up how much vitamin K₁ you are ingesting per day to see if you're getting enough—I can almost guarantee you are. That's because, as easy as phylloquinone is to get from food, blood clotting is too important to be left to the whims of dietary intake; the body needs to ensure it always has enough of this nutrient. The solution to a potential shortage isn't storage, either. Having a large reserve of K₁ hanging around might cause as many problems as not having enough, so it isn't stored in the body in appreciable amounts. Instead, the body has a special protective mechanism that recycles vitamin K₁ so dietary requirements are minimal and the vitamin is always available when needed, as the diagram below illustrates.

Since K₁ is required to ensure that blood clots appropriately, it's easy to predict and observe the symptoms of K₁ deficiency: bleeding. This may manifest as prolonged bleeding from a small wound, nose bleeds, bleeding gums, heavy menstrual periods and/or easy bruising. Because

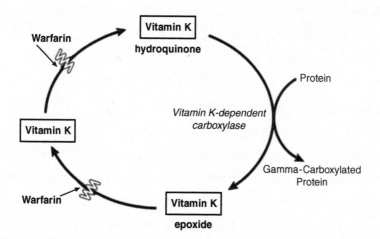

Recycling of vitamin K₁ in the body

the vitamin K_1 cycle safety net keeps dietary requirements to an absolute minimum, vitamin K_1 deficiency is rarely due to inadequate dietary intake. It is more often due to some other underlying medical condition, such as intestinal malabsorption syndrome or liver disease. Be sure to consult a health care professional if you experience any symptoms of insufficient blood clotting.

Vitamin K_2: Animal Fat and Calcium Metabolism

Like male and female fraternal twins, K_2 is as different from K_1 as two closely related individuals can be. Menaquinone—K_2—has very little to do with blood clotting; its job is to move calcium around the body. As explained in Chapter 1, K_2 activates certain proteins that guide calcium into bones and teeth where it belongs, and other proteins that escort calcium out of soft tissues, like arteries, where it is potentially harmful. This important function protects us against dental cavities, osteoporosis, heart disease, cancer and many other common ailments. Without K_2, calcium builds up in body tissues, where it is harmful, and fails to reach those areas where it is helpful.

Our menaquinone intake comes from two sources, diet and bacterial synthesis. As strange as the latter sounds, a very small amount of vitamin K_2 is made in the intestinal tract from dietary K_1 by the healthy bacteria that are normally present there. Unfortunately, this miniscule amount is not enough to prevent a vitamin K_2 deficiency if there aren't also any dietary sources. The level of K_2 produced by gut bacteria varies from person to person, and it is likely that none is produced at all in a person who has a history of antibiotic use or a condition that disrupts the natural gut flora.

The transformation of K_1 to K_2 varies per species. In ruminants, like cows and goats, as well as in other animals that are primarily herbivores, the conversion seems to happen readily. It makes sense that if an animal is well adapted to primarily eating green plants, its body will have evolved to extract or convert all the necessary nutrients from that food source. This is not the case with humans. Homo sapiens seem to convert very little K_1 to K_2, possibly because we evolved to a place in the food chain that provided us with ample sources of preconverted K_2, so we lost the ability to make it ourselves. As you'll see in the next chapter, those ample sources of premade K_2 primarily take the form of fat from animals who do convert K_1 to K_2 with ease. As a result, food sources of vitamin K_2 are very different from food sources of K_1.

Vitamin K₂ Deficiency Is Common and Invisible

Like K_1, K_2 is not stored in the body in significant amounts. Small amounts can be found in the salivary glands, pancreas, brain and sternum (the long, flat piece of cartilage that connects the ribs together in the middle of the chest). However, these stores are quickly depleted and, unlike K_1, *vitamin K_2 is not recycled.* That is why, as studies have shown, *humans can develop a K_2 deficiency in as little as seven days on a vitamin K–deficient diet,* which is one major factor in why inadequate vitamin K_2 levels are so common.[11] The other major reason for widespread K_2 deficiency is that getting it in our diet has become extremely difficult with modern foods. The next chapter, Chapter 3, is entirely devoted to explaining and solving this predicament, while Chapter 4 sheds light on why we have a mechanism to recycle one form of vitamin K but not the other.

Unlike the obvious bruising and bleeding caused by phylloquinone deficiency, a lack of menaquinone presents in a much more insidious fashion. Gradually waning bone density might progress for years before it is diagnosed as osteopenia or osteoporosis. Similarly, the first sign of calcium plaque obstructing the arteries to the heart might be a (possibly fatal) heart attack. Dental cavities and crooked teeth requiring braces are viewed as somewhat inevitable aspects of childhood—or at most, in the case of cavities, a lack of dental hygiene—rather than a nutrient deficiency. Until vitamin K_2 testing becomes a routine part of your annual checkup, as is now the case (or should be) with vitamin D, we all need to understand how to optimize K_2 intake.

One final note about this sibling pseudo-rivalry: the physiological actions of vitamins K_1 and K_2 aren't entirely mutually exclusive. There is a minimal amount of overlap that is likely contributing to the ongoing confusion around these fraternal twins. For example, studies show K_1 has minor bone-boosting ability, though we don't see any benefit from K_1 for heart disease.[12] Similarly, K_2 has a slight blood clotting action, but only at high intake levels. We'll get into the safety of K_2 and blood-thinning drugs in the next chapter.

The table below summarizes the similarities and differences between phylloquinone and menaquinone.

Similarities and differences between K_1 and K_2

	K_1 (phylloquinone)	K_2 (menaquinone)
Physiological action	Blood clotting	Appropriate calcification
Food sources	Green leafy vegetables, kiwifruit, vegetable oils	Natto (fermented soybeans); goose liver; certain cheeses; animal fat such as egg yolk, butter and lard of grass-fed animals

Stored in body?	No	No
Recycled in body?	Yes (so dietary requirements are minimal)	No (so dietary intake is crucial)
Deficiency	Uncommon—leads to bleeding disorder	Common—manifests as osteoporosis, arterial plaques, dental cavities
Ability to activate osteocalcin to build bone density and reduce hip fracture?	Slight	Strong
Ability to activate matrix gla protein (MGP) to prevent and reverse arterial plaques?	Slight	Strong

Vitamin K₂: A New Essential Nutrient

An essential nutrient, as mentioned earlier, is a nutritional element (such as a vitamin, mineral, amino acid or fatty acid) required for normal body functioning that either cannot be synthesized by the body at all or cannot be made in amounts required for good health and therefore must be obtained from a dietary source. Previous thinking on vitamin K₂ (for those who were thinking about it at all) was that, since gut bacteria can theoretically make menaquinone and the body's requirement for it is relatively low, it is not an essential nutrient. However, since we now know that the production by intestinal flora is inadequate and studies show that most people don't have enough K₂ to activate all of their menaquinone-dependent proteins, it is clear that K₂ must be obtained through diet or supplements and should definitely be classified as essential.

Now that you know the history of vitamin K₂, it is easier to understand how we overlooked this essential nutrient for so long. That being said, appreciating the differences between the two main forms of vitamin K only partially clarifies why we are much more at risk for a deficiency of K_2 than of K_1. Exactly what went wrong that landed us in our current K_2-deficient situation? This question we'll explore next as we look at the radical change to our food system that made menaquinone so hard to come by, and at how to get vitamin K_2 back into our bodies through food and supplements.

_3_

How Much Vitamin K₂ Do We Need, and How Do We Get It?

People of the past obtained a substance that modern generations do not have.

—Weston A. Price

I t seems rather suspicious to say, as I have, that almost everyone is deficient in vitamin K_2, doesn't it? After all, since a vitamin deficiency depends on dietary intake, its prevalence should vary within a population of individuals who have varying diets. And in theory it does—unless something happens to the food supply of the entire population that makes it almost impossible to get that nutrient. And that has been the fate of vitamin K_2. Menaquinone was once abundant in our diet. In our efforts to modernize food production for higher yields from smaller areas, we inadvertently eliminated this critical nutrient from our diet. In this chapter you'll learn how to get menaquinone back onto your plate, how much you need for optimal well-being and how to distinguish between the many K_2 supplements on the market.

The Lack of Vitamin K_2 in Our Diet: What Went Wrong?

Humans began domesticating animals between 4,000 and 10,000 years ago, depending on the area of the world you consider. In an evolution from the hunter-gatherer lifestyle, our ancestors realized that by managing the whereabouts of select docile creatures, we could benefit from a reliable source of nutrition without all the running about. The practice was simple enough: restrict the animals' roaming to a reasonable-sized area of their natural habitat (an area that provided said animals with ample food and water) and protect them from predators, and the animals will, in turn, provide us with dietary protein, essential fats, vitamins and minerals.

What is common to farming and hunting—the reason we consume animal-origin foods at all—is that it allows humans to profit, in a nutritional sense, from the sun's energy and the soil's minerals. These elements are captured by photosynthetic plants, then consumed and metabolized by creatures that are capable of doing so efficiently. During this process, animals conveniently transform the nutrients in these plants into forms that are more bioavailable to us humans. The nutrient content of meat, eggs and dairy products is a direct result of the composition of the animals' diet. In a very real sense, then, we are not just what we eat; we are also what our animals eat.

We have a deeply ingrained, and now largely inaccurate, idea of how this domestication thing works—the notion of cows grazing in a meadow. Indeed, if we look back as little as 100 years we would find that the vast majority of livestock did roam freely on green pastures. Times have changed. In 1800, about 95 percent of the North American population was rural, farming was nearly everyone's business and most families produced their own food.

By 1920, the rural population of North America had dropped to about 50 percent, and today less than 5 percent of our population makes their living in agriculture. The biggest change is that we are now dependent on a largely centralized, industrialized food supply. Even the folks who do make their living by ranching and farming generally do not produce their own food. Central to the industrialization of our food supply was the removal of livestock from the pasture and the invention of factory farming.

It was the discovery of vitamins A and D in the early 1920s that opened the door for large-scale grain feeding of livestock, the almost exclusive mode of commercial livestock management today. Adding these specific nutrients to feed meant that cattle, poultry and swine could survive without sunlight, a source of vitamin D, and without green grass, a source of nutrients from which animals can derive vitamin A.

This meant that animals could stay permanently indoors. But, despite that exclusive grain feeding was technically possible at this time, it still wasn't practical. Grain was expensive and grazing land was relatively cheap. That changed during World War II.

In the early 1940s, agricultural equipment manufacturers perfected the lightweight, self-propelled combine harvester. This grain-harvesting machine enabled farmers to produce much more grain than the nation's population could consume, and the price of grain plummeted as a result. Although it had long been common knowledge that providing some carbohydrate-rich grain, especially corn, would help fatten cattle that otherwise existed on green plants, until this time, grain feeding was the exception rather than the rule. With the emergence of the combine harvester, feedlots—also known as confined feeding operations or CFOs—were created, and factory farming was born.

In North America, the trend began in the beef industry. Texas, a state founded on ranching with grazing land aplenty, opened its first feedlot in 1950. The poultry industry began moving chickens off pasture and into buildings later that decade. The dairy industry followed suit in the 1960s, and pork producers did the same in the 1970s. Today the vast majority of North American poultry, eggs, meat and dairy are produced using confined, intensive farming techniques and grain feeding. Meat and egg marketers even promote the virtues of 100 percent grain-fed products.

So what? Well, *when we removed animals from the pasture, we inadvertently removed vitamin K₂ from our diet.* Remember when I said that humans can't really convert the vitamin K₁ from plants into vitamin K₂? Animals can—if they have abundant K₁ in their diets to begin with. Grain contains only a fraction of the necessary K₂ precursor found in green grass. When animals grazed on pasture, vitamin K₂ was abundant in our food supply. The most common dietary staples, like butter, eggs, cheese and meat, even when eaten in relatively small amounts, easily met our menaquinone needs. Now we consume large quantities of the mass-produced versions of these foods, but we are starving for the nutrients that they no longer contain.

The Grass-fed Vitamin

Dr. Weston Price, the dentist who discovered vitamin K_2, which he termed "activator X" (discussed in Chapter 2), noticed and then clearly demonstrated the relationship between grass feeding and vitamin K_2 content. He collected samples of dairy products every two weeks from multiple regions of the United States, Canada, Australia and New Zealand. Over the course of several years, Price analyzed more than 20,000 samples and noticed a very specific trend in the activator X content of butter samples that varied with the quality of cattle fodder. He concluded, "The factor most potent was found to be the pasture fodder of the dairy animals. Rapidly growing grass, green or rapidly dried [to preserve the green color], was most efficient" to produce activator X.[1] Price showed that both the activator X and vitamin A content of diary samples increased in the warmer months—usually with peaks in spring and fall that coincided with periods of rapid grass growth—and plummeted in the winter when cattle were consuming mostly dried (non-green) fodder.

The reason for this seasonal variation, and the reason grain-fed animal foods are lacking in K_2, has to do with the intimate relationship between vitamin K and chlorophyll, the pigment that makes green plants green. Vitamin K_1 is abundant in the membrane of the chloroplast, the part of a plant cell that captures sunlight for photosynthesis.

When cows, chickens or pigs consume green, chlorophyll-containing plants, they ingest phylloquinone (K_1), which is then converted to menaquinone (K_2). Grazing animals accumulate vitamin K_2 in their tissues in direct proportion to the amount of K_1 in their diet.[2] The lack of chlorophyll in grains means little K_1 and little or no K_2 in grain-fed animal foods.

The K_2-cholorophyll connection is also responsible for a unique characteristic of the fat of bona fide grass-fed foods: a distinct sunny yellow or orange tinge. Vitamin K_1 in green plants is almost always present alongside (but entirely distinct from) beta-carotene, another chlorophyll nutrient. Beta-carotene is the pigment that imparts an orange color to fruit and vegetables. Carrots, for example, are famous for their high beta-carotene content. Beta-carotene is also abundant in green plants—the yellow tone is just disguised by other pigments. When animals consuming green plants convert K_1 to K_2, beta-carotene hitches a ride. So, the fat of grass-fed animals is high in menaquinone and tends to have a more intense yellow or orange hue than the fat of non-grass-fed animals. This is a good rule of thumb to follow when selecting K_2-rich foods: in general, the more yellow or orange the fat, the higher the K_2 content.

We are now so used to ultra-white fat in our meat and poultry that it is even prized over yellow fat, which seems suspicious or unappetizing. Recognizing and appreciating the golden tones of pastured fat ("pastured" refers to a grass-fed product—not to be confused with "pasteurized," the process of partial sterilization by heat or irradiation) is an important step in the nutritional revolution that will get vitamin K_2 back onto our plates. Even the trend among some producers of grass-fed cattle to fatten them by grain feeding for a few weeks before slaughter quickly defeats the purpose of grazing. For example, because grain is more expensive in New Zealand than it is in North America, meat from grass-fed animals is still the norm, and animal fat tends to be more yellow as a result. In the mid-1990s, New Zealand beef producers experimented with taking cattle off pasture and fattening them American-style on grain to meet the Japanese demand for meat with very white fat. The experiment was a failure on many levels. Six weeks of grain feeding was not enough to completely remove the yellow tone to satisfy the Japanese market, but it did deplete the beta-carotene content (and presumably the vitamin K_2 and omega-3 content) by about 97 percent.[3] Grain feeding, even for a short period, drastically reduces the nutrient content of meat.

Another grass-fed source of protein is wild game. Duck, pheasant, rabbit, venison, elk, boar, wild turkey and so on naturally thrive on green vegetation. A drawback here is that wild game tends to be quite lean, so the overall content of fat necessary to provide fat-soluble vitamins is low. If you are fortunate enough to have access to nature's bounty in this way, don't waste any of the precious, menaquinone-rich fat wild game provides.

Although total grass feeding of farmed animals will restore the maximum amount of K_2 to our meat, milk and eggs, it doesn't take much grazing to enhance the nutritional value of butter. Fresh grass in the cow's diet improves the nutrient content and flavor of butter in direct proportion to the amount of green fodder she takes in.[4] Just a few hours of grazing per day is better than no grazing at all, and it produces measurable improvements in the nutritional quality of dairy products. We'd be taking a step forward if meat and dairy industry practices progressed to provide old-fashioned grass or pasture access for cows, chickens and pigs.

Industry supporters of factory farming might argue that grazing cattle takes up too much land and that grass-fed butter is soft and difficult to ship, making it impractical to feed the masses (see the sidebar "How spreadable is your butter?" in Chapter 4). Granted, golden orange, 100 percent grass-fed "June butter" might be the holy grail of pastured products, but it is darn hard to come by. For readers who want to start reaping the artery-clearing, bone-building benefits of grass-fed foods as soon as possible, there are other options. In particular, one potentially pastured product is much more accessible because it can be produced in relatively small areas and ships easily, even when loaded with menaquinone-rich, grass-fed goodness. It is the humble egg. This kitchen staple already lies at the heart of a grassroots (no pun intended) movement that will help restore our depleted K₂ status.

The Pastured Egg Hunt

If you have only ever eaten factory eggs, this concept may be foreign to you, but the quality of an egg varies widely depending on what the chicken ate. For those fortunate to have had the pleasure of eating pastured eggs, you'll know what I mean. The white of an egg from a grass-fed hen is firmer and not as watery as that of an industrial egg. The yolk is always much deeper and darker in color than a conventional egg. At certain times of the year (there's a natural, seasonal variation), this color reaches a startling shade of deep, golden orange. That's literally nutritional gold—and you can expect it to be very high in heart-healthy K₂.

If you want to imagine just how orange a pastured egg yolk can get, picture cheddar cheese or, more specifically, Kraft macaroni-and-cheese powder. Its intense, glowing orange is often mocked for being the epitome of synthetic—and I'm not trying to suggest otherwise—but its instinctive appeal to the preschool set might just be because we are programmed to appreciate this color on some level. Perhaps it's because, in nature, this shade of orange is associated with nutrient-dense food. On the few occasions I gave in to my son's pleas for "the orange noodles" and served them, I snuck a whole grass-fed egg yolk into his serving. The color was indistinguishable and the suspicious side dish was transformed into a delivery system for (cleverly disguised) vitamin K_2.

If conventional grain feeding reduces the menaquinone content of egg yolks, what's a better egg, at least of those available at your local supermarket? Like me, you might be overwhelmed by the selection of eggs these days. There are the regular eggs that, last time I looked, are organized by size, color and brand. These are the standard, mass-produced eggs from never-see-the-light-of-day-much-less-a-grassy-field, battery-cage chickens. Then there are the "healthy" eggs, often found in a separate area of the store, which come in an increasingly baffling array of choices. These include "cage-free," "free-range," "free-run," "organic," "natural," "omega-3" or any combination of these terms. Which of these is the best? As far as K_2 is concerned, it doesn't make a bit of difference. If your "free-range" chicken isn't spending any time on green pasture, menaquinone content will be minimal.

As discouraging as this sounds, it's not the end of the story. The diverse supply of alternative ova reflects the growing demand for a better egg. It also hints at a phenomenon that's exploding outside the grocery store: a massive quest for farm-fresh eggs laid by chickens that scratch and wander on pasture. Foodies have been desperately seeking the elusive, truly free-range egg for a few years now because of its reputedly superior flavor. In some areas, alternative eggs have reached a cult status, making stars of the farmers (or starmers) who raise them. This has even created a gray market for the perfect egg, which is invariably offered by local producers whose small flocks are not subject to industry regulations. Top this off with the rapidly growing trend of urbanites and suburbanites keeping hens in their backyards and you have a veritable mania for eggs from grass-fed chickens. These, whether you locavores knew it or not, are the eggs that will help restore our K_2 status.

I confess, I have been buying my pastured eggs off the gray market, since local bylaws stubbornly prohibit me from having my own little flock of chickens. What do I do on the frequent occasions that my local egg lady is fresh out? I visit my local health food store for Vita Eggs, a Manitoba-based brand that supplies those lovely orange-yolked eggs almost year-round. The yolk color isn't due to grass feeding (sigh) but to the additions of herbs to the diet of these cage-free chickens. Still, the fact that the birds are getting some greens in their diet should augment the K_2 content of the yolks; indeed, their telltale orange tone proves it. Vita Eggs are an encouraging sign that a more nutritious egg can be produced on a large scale.

A good egg

A direct comparison of K_2 content between conventional eggs and grass-fed eggs has yet to be completed, but there are many other proven nutritional advantages to pastured eggs. Compared with official nutrient data for commercial eggs, eggs from hens raised on pasture have:[5]

☐ one-third less cholesterol
☐ one-quarter less saturated fat
☐ two-thirds more vitamin A
☐ two times more omega-3 fatty acids
☐ three times more vitamin E

- □ seven times more beta-carotene
- □ 50 percent more folic acid
- □ 70 percent more vitamin B$_{12}$[6]
- □ four to six times more vitamin D

As you will learn in Chapter 7, it's ironic that the discovery of vitamins A and D paved the way for eliminating K$_2$ from our diets, given the intimate relationship between fat-soluble vitamins. But since we didn't know K$_2$ was there in the first place, we didn't know we were doing away with it. Why doesn't the loss of vitamin K$_2$ limit the survival of confined animals, the way a vitamin A or D deficiency would? As with humans, life-threatening problems with K$_2$ deficiency in animals tend to manifest in the long term. In the artificially short lives of factory-farmed creatures, a menaquinone deficiency can go unnoticed. For example, a broiler chicken raised on pasture takes at least three months to mature, whereas a confined grain- (and usually hormone- and antibiotic-) fed bird takes a mere seven weeks. Industrial farming has taken fast food to a new level, and we've paid the price with nutrient deficiencies.

The demand for pastured foods has been sparked by, in addition to a desire for improved flavor and nutrition, an increased awareness of the often deplorable living conditions of conventional livestock and of the environmental impact of industrial farming. A discussion of the many ethical, environmental and full nutritional implications of factory farming is beyond the scope of this book, as is an investigation into the logistics of providing pastured animal foods for a planet of 7 billion people. But to make a long story short, what's good for our animals is good for us. Finding products from grass-fed sources often takes extra effort, but it's an important step in restoring vitamin K$_2$ and many other nutrients to our diet.

Choosing dairy products from smaller, local or organic dairies will boost your chances, although not guarantee, that the cows are getting some grass in their diets during the summer. An organic designation itself means nothing when it comes to grass feeding. However, small and organic dairies are more likely to collect milk from a number of independent dairy farms that may be letting their cows graze. If they are, then you'll get more K$_2$ in your full-fat milk, yogurt and butter. You can check the dairy's website or make inquiries about the cows' fodder if you want to know more about the quality of your dairy products. Don't take it for granted that the cows are spending any time on pasture, even if the company's logo features an artist's rendition of a cow in a field.

Beef from grass-fed cows

In addition to having a higher K$_2$ content, grass-fed beef is better for human health than grain-fed beef in 10 ways, according to the most comprehensive analysis to date.[7] Compared with grain-fed beef, grass-fed beef is:

- ☐ lower in total fat
- ☐ higher in beta-carotene (linked to K$_2$ content)
- ☐ higher in vitamin E (gamma-tocopherol)
- ☐ higher in the B vitamins thiamin and riboflavin
- ☐ higher in the minerals calcium, magnesium and potassium
- ☐ higher in total omega-3 essential fatty acids

- ☐ higher in conjugated linoleic acid (CLA), a healthy fat and potential cancer fighter
- ☐ higher in vaccenic acid (which can be transformed into CLA)
- ☐ lower in the saturated fats

It also has a healthier ratio of omega-6 to omega-3 fatty acids (1.65 to 4.84).

Ghee and Butter Oil from Grass-Fed Cows

One pastured dairy product that is available to everyone at health food stores or by mail order is ghee made from the milk of grass-fed cows. Ghee is also known as Indian clarified butter, drawn butter, butter ghee or anhydrous milk fat. This traditional food is prepared by melting and simmering unsalted butter over a low temperature until the water evaporates and milk solids separate from the oil. The resulting oil has a semisolid consistency and is very stable. It can be stored without refrigeration for as long as three months, and up to a year in the fridge. Not all ghee comes from grass-fed cows, however, so look for the telltale golden yellow color and read the label carefully to make sure you are getting the most menaquinone bang for your butter buck.

Ghee made from the milk of grass-fed cows certainly fits the description of the activator X–rich golden butter oil that Dr. Price described in his work. Based on the green content of the cows' fodder and the careful concentration of the fat (and therefore fat-soluble-vitamin–containing) portion of the butter, pastured ghee should be a plentiful source of vitamin K_2. Ghee can be used wherever you would use butter, and its high smoke point (485°F) makes it a great choice for frying and sautéing.

One company that produces an excellent pastured ghee is Pure Indian Foods. Its label states that "this ghee is made with milk obtained only during the spring and fall, when the cows are out to pasture eating rapidly growing green grass." As a bonus for followers of ayurvedic medicine, a traditional system of medicine in India ("ayurvedic" is Sanskrit for "the complete knowledge for long life"), this product is made according to Vedic principles. The label also explains, "We make our ghee only during the full or waxing moon days," which, apparently, is good news for Vedic followers.

I don't know to what extent the waxing of the moon contributes to the quality of the end product, but I have to say, this stuff is absolutely delicious. I was blown away by the flavor of this grass-fed butter ghee, a taste that I can only describe as, well . . . buttery. If you have ever eaten artificial butter–flavored popcorn and wondered who decided that *that* flavor represented actual butter, you'll have an "aha" moment when you taste grass-fed ghee. It captures all the best butter flavor without the phony odor (or questionable radiation) of microwave popcorn. To boot, it should be chock-full of K_2. Unfortunately, at about $1 per ounce—plus shipping, which may cost as much as the product itself depending on where you live—grass-fed ghee is a luxury item for most people.

Also in the category of grass-fed butter concentrates is butter oil. The most prominent brand is X-Factor Gold High Vitamin Butter Oil, produced by Green Pasture Products and sold by several online retailers. According to the label, it is, like grass-fed ghee, made from dairy oil extracted from cows that eat "100 percent rapidly growing grass." The product label further explains that "the speed of grass growth, timing of the grazing, species of grass, climate and extraction method are all important" in producing the butter oil.

Although X-Factor butter oil is marketed as a dietary supplement, I would put it in the same "functional food" category as grass-fed ghee, because the nutrient content is not listed on the label. However, since the butter oil is *seven times* the price per ounce of grass-fed ghee, I did endeavor to determine exactly what distinguishes it. Except for the reference to an unstated species of grass on the butter oil label, the label descriptions of the two products seem the same. I interviewed the owner of Green Pasture Products and my specific questions were met with vague answers. If you can afford this product, you can afford grass-fed ghee, and, according to my research, you'll be getting pretty much the same thing.

Yet Another Reason to Avoid Trans Fat

In addition to the gradual loss of foods from grass-fed sources, our K_2 status took another collective hit with the advent of trans fat. Simply put, *eating processed or fast food increases our risk of vitamin K_2 deficiency.* More than just displacing butter—which, grass-fed or not, at least stands a chance of containing some menaquinone—margarine and other hydrogenated oils deliver a sucker punch to our K_2 intake. These butter substitutes introduced a mutant form of vitamin K called dihydrophylloquinone (DHP) into our diets. DHP is formed when vitamin K_1–rich plant oils are synthetically hydrogenated. Commercially baked goods and fried foods are major dietary sources of DHP, and blood levels of this antinutrient are used in scientific studies as a marker for low-quality diets.

What does this have to do with K_2? Well, even when you adjust for other markers of diet quality such as calcium intake, and relevant lifestyle factors such as age, body weight, exercise and estrogen use, higher DHP intake is associated with lower bone mineral density in both men and women.[8] In other words, the ill effects of this synthetic form of K_1 spill over into the bone density domain of K_2. That is probably because the meager conversion of natural K_1 to K_2 happens even less with DHP. Studies show that vitamin K_2 tissue concentrations are much lower when DHP is the only dietary source of K_1, versus diets containing natural K_1.[9]

We know that snacking on DHP-laden muffins and french fries causes K_2 levels and bone health to suffer, but what about heart health? It is well established that the hydrogenated fats invented to replace saturated fat ironically turned out to be much worse for heart health than old-fashioned butter. However, the standard explanation for this effect is that trans fat increases systemic inflammation and raises LDL, the so-called bad cholesterol. In fact, another, more directly harmful factor is at play here. Trans fat, even in small amounts, increases the incorporation of calcium into atherosclerotic plaques.[10]

Several mechanisms chip in to trans fat's negative impact on heart health, but its effect on calcium plaque formation has everything to do with the fact that trans fat contains the vitamin K deviant, DHP. It is highly likely that since DHP is unable to activate the protein osteocalcin, it is equally ineffective at activating calcium-clearing MGP (matrix gla protein). That leaves us prone to heart disease as well as wrinkles, varicose veins and the other conditions we'll cover in Chapters 4 and 5. If you aren't already doing so, steer clear of hydrogenated fats. In addition to reading the nutrition information labels on food packages, scan ingredient lists. Current label laws in both Canada and the United States allow for any food containing less than one-half gram of trans fat to be labeled as trans fat–free, which is misleading to consumers. If the words "hydrogenated" or "partially hydrogenated" appear on the label, the contents contain trans fat. Likewise, the terms "mono-glycerides" and "di-glycerides" should also tip you off to the presence of Frankenfat in the food. By avoiding DHP, you'll be giving K_2 a helping hand.

Two Types of Vitamin K_2

I know what you're thinking: Okay, I can avoid trans fat, but what if I can't find pastured eggs and dairy products? Am I doomed to a life of wrinkles and varicose veins? No, you are not. While we are waiting for the supply of foods from grass-fed sources to catch up to the demand, there are other food sources of menaquinone that will help satisfy our need for K_2. Before we delve into what those sources are, it is useful to know about two main forms of menaquinone found in food. They both provide the benefits we seek from K_2, but they are found in different types of food.

In Chapter 2 I mentioned that we get vitamin K_2 from two sources, diet and intestinal bacteria, but that the latter contributes only a negligible amount to our K_2 status. In other words, the amount of K_2 synthesized by bacteria in the human intestines won't save humans from K_2 deficiency. However, there are other microorganisms in nature that produce K_2 very efficiently, providing us with menaquinone-rich foods, namely certain types of cheeses and a Japanese soy food called natto. The two main food sources of K_2—animal and bacterial—each supply a different type of menaquinone.

The kind of K_2 synthesized by mammals and found in grass-fed meat, egg yolks and butter is called menaquinone-4, or MK-4 for short. It is so named because the molecular structure of this form of K_2 has a hydrocarbon "tail" that contains four double bonds. Bacterial fermentation, on the other hand, produces a range of other menaquinones, designated MK-5 to MK-10, depending on the specific microorganism in the food. Of these, menaquinone-7, or MK-7, is especially important. Its structural tail contains seven double bonds, and it is the primary menaquinone found in the superfood natto. Don't bother stretching your brain back to high school chemistry to remember the significance of double bonds. Just know that the structures of MK-4 (animal origin) and MK-7 (bacterial origin) vary slightly and this structural variance confers different properties to each form of menaquinone. These properties will become more important when we discuss choosing a supplement later in this chapter. The health benefits are the same for each type when taken in appropriate doses.

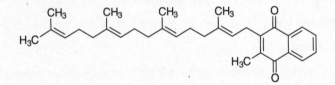

Molecular structure of menaquinone-4 (MK-4)

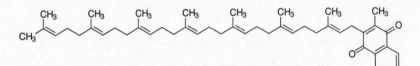

Molecular structure of menaquinone-7 (MK-7)

Vitamin K₂ content of selected foods

Food (3 1/2 ounce portion)	Micrograms	Proportion of vitamin Ks
Natto	1,103.4	(90% MK-7, 10% other MK)
Goose liver pâté	369.0	(100% MK-4)
Hard cheeses (Dutch Gouda style)	76.3	(6% MK-4, 94% other MK)

Soft cheeses (French Brie style)	56.5	(6.5% MK-4, 93.5% other MK)
Egg yolk (Netherlands)	32.1	(98% MK-4, 2% other MK)
Goose leg	31.0	(100% MK-4)
Egg yolk (U.S.)	15.5	(100% MK-4)
Butter	15.0	(100% MK-4)
Chicken liver (raw)	14.1	(100% MK-4)
Chicken liver (pan-fried)	12.6	(100% MK-4)
Cheddar cheese (U.S.)	10.2	(6% MK-4, 94% other MK)
Meat franks	9.8	(100% MK-4)
Chicken breast	8.9	(100% MK-4)
Chicken leg	8.5	(100% MK-4)
Ground beef (medium fat)	8.1	(100% MK-4)
Chicken liver (braised)	6.7	(100% MK-4)
Hot dog	5.7	(100% MK-4)
Bacon	5.6	(100% MK-4)
Calf's liver (pan-fried)	6.0	(100% MK-4)
Sauerkraut	4.8	(100% mixed MK)
Whole milk	1.0	(100% MK-4)
Salmon (Alasksa, Coho, Sockeye, Chum and King wild (raw))	0.5	(100% MK-4)
Cow's liver (pan-fried)	0.4	(100% MK-4)
Egg white	0.4	(100% MK-4)
Skim milk	0.0	

Sources: Schurgers LJ, Vermeer C. Determination of phylloquinone and menaquinones in food. Effect of food matrix on circulating vitamin K concentrations. *Haemostasis.* 2000 Nov-Dec, 30(6):298-307; Elder SJ, Haytowitz DB, Howe J, et al. Vitamin K content of meat, dairy and fast food in the U.S. diet. *J Agric Food Chem* 2006, 54:463–67.

Now that you know about the different types of K₂, you can understand why some foods, like cheese, are happily higher in menaquinone than you might expect. Whole milk from your average grain-fed cow contains a paltry 1.0 micrograms of K₂ per 100 milliliters. Cheese made from that milk can be as high as 76.3 micrograms per 100 grams. That's because bacterial fermentation augments the K_2 content. In this example only, about 6 percent of the menaquinone content is from the milk that went into the cheese; bacteria generously provided the rest. The "other MKs" (other menaquinones) in the list refer to MK-5 through MK-9, the range of long-chain forms of vitamin K_2 produced by bacteria. According to our current understanding, they all have the same health benefits.

Notice that egg yolks in the Netherlands contain more than double the K_2 of American (and presumably Canadian) egg yolks. That reflects feed quality and the fact that the average Dutch chicken is more likely than her North American counterpart to spend some time in the great outdoors. Another unmistakable trend is the very high K_2 content of rich, indulgent "sin" food like goose liver and fatty cheeses. This explodes the alleged contradiction between the indulgent consumption of rich food in European diets and the relatively low rate of heart disease in Europe. The French Paradox isn't a paradox at all. The very same pâté de foie gras, Camembert, egg yolks and creamy, buttery sauces that we inaccurately labeled "heart attack on a plate" liberally supply the single most important nutrient to protect heart health. Good news for the bon vivants among us.

On a less gourmet note is the K_2 content of frankfurters and hot dogs. I debated long and hard as to whether these entries should be included in the list. I'm not trying to encourage people to rely on or justify eating these highly processed, nitrate-laden items for their slight K_2 content. I ended up including them to make the point that the unexpected menaquinone content might very well reflect the higher amounts of K_2-rich organs that go into these "mystery meats."

Natto, the Slimy Superfood

Of all the menaquinone-containing foods, one unusual superstar tops the list. That is natto, a Japanese breakfast treat that smells like old gym socks, is held together by gobs of stringy mucus and contains enough K_2 per serving to prevent hip fracture and heart disease. It's hard to find and even harder to like, but if you can wrap your mind around blue cheese, natto shouldn't be such a stretch. What's more, there's finally some good news for vegans here: although in the Western diet, K_2 would traditionally have been obtained only from animal foods, the introduction of natto does offer one plant-based source of this nutrient.

It is not exactly clear when or how natto originated, although people in Japan have been enjoying the beans for several centuries, and many theories exist about its invention. According to one popular legend, Japanese soldiers inadvertently created natto sometime around the year 1080. As the story goes, the warriors were boiling soybeans for horse feed when an invading army suddenly attacked. The beans were hurriedly thrown into straw sacks, where they remained until the army returned to camp a few days later. By that time, the beans had rotted, but the desperate soldiers ate them anyway and (this is the suspicious part) liked the flavor.[11] Natto was soon brought to the reigning emperor, who also declared it delicious, and its popularity spread from there.

Unbeknownst to those hungry soldiers and their commander, the bacteria in the straw sacks that caused the soybeans to rot also happened to produce copious amounts of vitamin K_2. Until recently, natto was still prepared using the traditional methods of packing boiled soybeans in straw. Today, straw is no longer directly involved in natto production, but the specialized K_2-producing microbe from the straw,

Bacillus subtilis natto, is added to the beans in a factory. This results in a more consistent product that is still brimming with menaquinone.

Natto isn't enjoyed everywhere in Japan. It is frequently eaten in the eastern regions (near Tokyo) but seldom in the western part (near Hiroshima) of the country. As it turns out, regional differences in natto noshing have a major impact on hip fracture rates. Studies show a statistically significant inverse correlation between the incidence of hip fractures in women and natto consumption in each prefecture throughout Japan.[12] In other words, hip fracture rates are lower in areas where people eat natto. That's because both MK-7 and activated osteocalcin levels are higher in people who eat natto, even occasionally.[13] One two-ounce serving of natto provides a whopping 550 micrograms of bone-building, plaque-busting vitamin K_2 in the form of menaquinone-7.

Natto boasts other health benefits in addition to its superior K_2 content. Pyrazine, the compound that gives natto its distinct smell, reduces the likelihood of blood clotting, another bonus for cardiovascular health. Nattokinase, a protein-digesting enzyme unique to this fermented food, has also demonstrated some clot-busting activity and potential for preventing or treating Alzheimer's disease.[14] As well, natto contains vitamin PQQ (pyrroloquinoline quinone), a micronutrient with skin-health benefits that will probably become better known in the coming years. Like other soy foods, natto is a source of daidzein, genistein, isoflavones and phytoestrogen, compounds with alleged anticancer activity. There are many reasons to love natto, and I was determined to do just that while writing this book.

Unlike the passion for pastured eggs, a natto craze isn't exactly sweeping the nation, so finding the latter is even more challenging than sourcing the former. After reading much about the infamous stuff online, and watching a few amusing YouTube videos of people eating natto for the first time, I set out to buy some. Toronto is a multicultural city with a high Asian population, although admittedly not a strong Japanese presence. Even so, I thought natto would be a cinch to find. I was wrong. A search of several large, Toronto-area health food stores turned up nothing but shoulder shrugs from the staff. Ditto for a number of well-stocked Asian supermarkets. After being stumped for a few weeks, I eventually hit the jackpot by stumbling upon a Japanese restaurant that had natto rolls on the menu.

I inquired about the natto and asked if it was possible to order some to take home. The restaurant's Japanese owner and sushi chef, Simon, rushed over to my table, all smiles. He was obviously thrilled with my interest in natto and proceeded to go on about how great it is. He especially emphasized how healthy natto is, although he didn't seem clear on exactly why. Simon admitted he didn't always like natto, but after trying it a few times, he was converted. I had heard a similar story from a few Caucasians and I was encouraged to hear that natto is an acquired taste, even for the descendants of its inventors. Surely I could learn to love natto too. The restaurant had two containers of natto to spare and I placed an order of several more for the following week. I went home with my little white containers for which, I later learned, I dearly overpaid at $5 a pop. The following day I steamed some rice and opened my first container, thus officially entering the weird and not-so-wonderful world of natto.

Natto comes in a small, square Styrofoam pods that hold about an ounce and a half of ooey, gooey beans. They smell a lot like the kitchen compost bin when it's overdue for emptying. Although off-putting, the moldy food smell is relatively mild, especially when natto is eaten cold.

I found the nutty, savory, slightly salty and musty flavor to be mostly inoffensive as well, although this is likely one of those foods that individuals could experience very differently. The issue for me was the texture. I don't consider myself to be especially texture sensitive when it comes to food, but this was ridiculous.

Japanese have a specific word to describe the texture of natto: *neba-neba*. The closest word we have in English would be "gooey," although my understanding of *neba-neba* is that it connotes a certain element of stickiness within the sliminess. Natto novices might think, as I did, that the challenge is preparing the beans in a way that minimizes the slimy-sticky quality. It doesn't work that way. Nothing you do to natto will change or disguise its distinctive texture. In fact, anything you add to natto will assume the same stringy, slippery consistency. You just have to embrace it.

Preparing natto is easy. It comes pretty much ready to eat, although it might need to be thawed first in the handy, single-serving-size container. Miniscule packets of condiments such as soy sauce and Japanese mustard are even included in most brands. All you have to do is add the sauces, mix for a minute to "develop the spider's web," as Simon put it, and serve over rice. Minced chives or green onion is a standard garnish, as is—brace yourself—a raw egg. The latter is entirely optional, but it is a traditional and popular addition. This is breakfast, after all.

Natto is highly versatile. In addition to the standard natto over rice, these beans can be served in a multitude of ways, of which I tried several. There's natto on toast, natto spaghetti and natto fried rice. There's even natto ice cream, which I did not have the pleasure of sampling. Every few days I prepared natto in a different way until I found a recipe that enabled me to finish a whole, tiny serving. Along the way I learned an important lesson: don't scratch your forehead while eating natto. If any of those stringy filaments happen to be stuck to your hand, you'll end up feeling like you walked into a cobweb.

I have to admit that, ultimately, after considerable effort, I could not embrace the texture of natto. Perhaps I'm easily discouraged, but eating the fermented beans soon became a dreaded chore. I really wanted to be able to tell you that at some point I suddenly "got" natto the way I get blue cheese, but it just didn't happen. I reached a point where I could tolerate eating the stuff, but that's it. Even my "favorite" dish, natto fried rice (see the sidebar for the recipe), is just edible. I made a sincere attempt to love natto, and I failed, but please don't let that discourage you from trying. It is scientifically proven that you will reap the K_2-associated benefits if you succeed at loving it—or if you can at least choke it down once every couple of weeks, which is what I will continue to do.

Natto fried rice

1 packet natto, thawed

2 eggs, beaten

1 tbsp vegetable oil

1 tsp sesame oil

2 cups cooked white or brown rice

2 tbsp chopped scallions

Salt and pepper to taste

Soy sauce to taste

Stir the natto for 1 minute to thicken. In a bowl, combine natto with eggs and mix well. Heat stir-fry pan over medium-high heat and add vegetable oil. Tilt the pan so oil coats the surface evenly, then remove excess vegetable oil and reserve. Add sesame oil to pan. Add the egg and natto. Sauté, stirring frequently, until egg is fully cooked and set. Remove natto mixture and set aside. Return reserved vegetable oil to the pan, then the cooked rice. Sauté rice until hot, breaking up any lumps. Add scallions to rice and cook, stirring, for 1 minute. Return the natto mixture to the pan and mix well with the rice. Season to taste with salt, pepper and soy sauce. Remove from heat and serve immediately. Serves 1 to 2 people.

Does it have to be natto, you ask? Couldn't you eat less offensive forms of soybeans instead, like tofu, miso or soy milk? Nope. Granted, most soy foods are a source of isoflavones, plant-based estrogen-like compounds for which there is evidence (albeit disputed) of bone-boosting activity. But these soy products do not contain menaquinone, so they won't benefit your heart and blood vessels the way natto does. Studies confirm that eating natto is associated with improved bone density to a much greater extent than are other soy foods, such as tofu.[15]

Recommended Vitamin K₂ Intake

Now that we've examined which familiar and foreign foods provide menaquinone, you might be wondering how much K_2 you really need for optimal bone, heart and overall health. Establishing this provides a framework for choosing K_2 dietary supplements. You can disregard any K_2 prescriptions based on the current recommended daily intake (RDI). The official RDI for vitamin K doesn't distinguish between the two main forms of vitamin K, and it is based on the body's requirements for K_1, not K_2. The RDI for vitamin K is defined by the liver's requirement for normal clotting factor activation; it does not account for how much vitamin K₂ might be needed for optimal bone and artery health. As you learned in Chapter 1, blood clotting proteins are completely activated in most healthy people, but a varying portion of K₂-dependent proteins remain inactive in those same people. In other words, the current recommended intake of "vitamin K" leaves vitamin K₂-dependent proteins lacking in vitamin K₂.

A panel of European experts who convened in 2004 to discuss this very matter concluded, from the available scientific data, that considerably higher intake of vitamin K is required for optimal gamma-carboxylation (the process by which K₁ and K₂ activate proteins) of osteocalcin.[16] In other words, we need to establish new guidelines that reflect the vitamin K needed to meet our requirements for both blood clotting and protection from the Calcium Paradox. Current recommendations are based on levels that ensure adequate blood coagulation but fail to ensure long-term optimal levels of vitamin K₂, leaving us at risk of bone fragility, arterial and kidney calcification, cardiovascular disease and possibly cancer. Scientific research has demonstrated that markedly higher osteocalcin activation is obtained by intakes of vitamin K well above the current recommended dietary intake.[17] Chapter 4 explains how we get by being slightly or even severely vitamin K₂ deficient now by paying the price later in life.

You can expect scientific studies to confirm the specific dose of vitamin K₂ for optimal long-term health sometime after this book goes to print. In the meantime, we can glean practical information from the doses used in clinical trials and from population-based studies that monitor K₂ intake. Researchers see a reduction in arterial calcification and cardiovascular mortality with as little as 45 micrograms of vitamin

K_2 daily.[18] Frequent natto eaters may be getting more than 300 micrograms of menaquinone every day. Vitamin K_2 has no known toxicity, so we need to establish minimum and useful maximum intake levels. To a certain extent, an optimal dose of K_2 will depend on your intake of vitamins A and D, but we'll get to that later. For now just keep in mind that the target dose of K_2 depends on the type of menaquinone you consider. For this reason, I'll address them separately.

Vitamin K_2 from Supplements

What if, despite spending your weekends combing the local farmers' markets, you have been unable to source (or afford) the elusive foods from grass-fed sources that should be our birthright? What if you can eat only so much goose liver pâté or aged Dutch Gouda? And what if you just can't stomach natto? Then it's time to talk menaquinone supplements. Just as there are two main types of K_2, there are two major categories of K_2 supplements: MK-4 products and MK-7 products. It is critical to understand the difference between them to ensure you are getting what you paid for, and that you're taking an appropriate dose of the product you choose.

MK-4 Supplements

MK-4 supplements have been available in the United States for many years, and many scientific studies about the benefits of K_2 come from studies using these products. Although MK-4 is the natural form of K_2 found in animal foods, this is *not* the source of K_2 for supplements. Extracting MK-4 from grass-fed butter or egg yolks would be exorbitantly expensive. MK-4 in supplement form is synthetic, typically made from an extract of the plant *Nicotiana tabacum*, or common tobacco. This does not in any way liken supplementing with MK-4 to smoking cigarettes. MK-4 might be listed as "menatetrenone" on the label, and 45 milligrams daily is a typical therapeutic dose of MK-4 used in clinical research.

Whatever your feelings on synthetic versus natural supplements, clinical trials show that you can reap all the bone-building, artery-clearing benefits from this form of menaquinone. As such, I would not hesitate to recommend MK-4 products, if it weren't for two major drawbacks. First, MK-4 has a relatively short half-life. "Half-life" refers to the time required for a substance to decrease its concentration in the body by half. It is a measure of how long a given substance stays in the body. For example, if a nutrient (or drug or whatever is being measured) has a half-life of an hour, one hour after ingesting it, one-half of the original nutrient dose will be cleared out of circulation. After two hours, three-quarters of the nutrient will be depleted; after three hours, only one-eighth of the original substance will remain and so on. MK-4 stays in circulation for only a few hours before blood levels drop below a therapeutic amount. To maintain useful levels of synthetic MK-4, it must be taken throughout the day. The three-times-daily dosing is inconvenient and could result in people taking less than an optimal amount if doses are missed.

The second drawback of MK-4 applies primarily to supplement buyers in Canada, but it's an important one. Health Canada, which allowed vitamin K in any form to be legal for sale only as of 2005, has set a limit of 120 micrograms (0.120 milligrams) of vitamin K per daily dose for any product. This very conservative limit was set with vitamin K₁ in mind, to limit the availability of the form of vitamin K that might interfere with blood clotting. However, 120 micrograms is only 2.6 percent of the recommended 45 milligram (4,500 microgram) dose of MK-4. No studies have even been published using MK-4 at this low dose, so there is no scientific evidence to suggest it is therapeutic. Until Health Canada differentiates between allowable K_1 and K_2 levels, and drastically increases the allowable amount of K_2, MK-4 products in Canada are essentially useless. Fortunately, consumers on both sides of the border have another option.

MK-7 Supplements

MK-7 products are relative newcomers to the K_2 supplements scene. These are sourced from natto, and early evidence shows they are just as effective as MK-4 products at protecting your heart and bones, with at least one big advantage: convenience. MK-7 has a longer half-life in the body, so a single daily dose provides continual K_2 protection.[19] MK-7 supplements provide higher and more stable menaquinone blood levels than MK-4 products.

Another advantage of MK-7 products is that an effective daily dose is approximately 120 micrograms. Milligrams or micrograms, it all fits into a pretty small capsule, so why is that such a big deal? Remember, 120 micrograms is the maximum daily dose you'll find in any product on Canadian shelves. Fortunately, MK-7 allows us to reap all the health benefits of vitamin K_2 within our relatively restricted daily intake.

By the way, fellow Canadians, just because Health Canada limits the daily dose of K_2 available in a supplement of K_2 to 120 micrograms doesn't mean you can't take more. Pesky restrictions aside, vitamin K_2 has no known toxicity, so you can safely double up on your dose. Eating a single 40 gram serving of natto every day provides more than

400 micrograms of K_2, if you can stand it. Furthermore, as I explain in Chapter 4, studies show that menopausal and postmenopausal women have a higher need for K_2, so taking 240 micrograms or more of MK-7 daily is a particularly good idea for women who are facing the big change.

If the label doesn't specify, not even in the fine print, exactly what type of vitamin K is in the supplement, don't buy it. If you can't tell what type of K you are buying, you can't know if the dose is appropriate—don't assume the manufacturer got it right. On that note, I wouldn't bother with a product that contains vitamin K_1. This advice applies especially to Canadian consumers, who are limited to a minimum amount of K_2 per daily dose. Vitamin K_1 is so easy to get in our diet, the body recycles it and deficiency is rare, so why pay to have it in your supplement and sacrifice your allowance of precious K_2?

One caution when choosing a K_2 supplement: since MK-7 is extracted from natto, it may pose a problem for those who are allergic to soy, so MK-4 would be the best choice in that case.

Cross-Border Shopping

Watch out for brands of MK-4 products available in both Canada and the United States. A handful of natural health product manufacturers that were producing K_2 products for the U.S. market prior to 2005 simply kept using the same MK-4 material when Health Canada gave vitamin K supplements the green light. South of the border, these K_2 products recommend a standard 45 milligram (45,000 microgram) dose of MK-4, whereas north of the border, a product of the same brand contains only 0.120 milligrams (120 micrograms) of MK-4 per daily dose, as per Canadian regulations. I have not been able to get a convincing answer as to why the "same" product is almost 40 times more potent in one country than another, nor how the same claims of efficacy can be made about products with only 3 percent of a therapeutic dose. Canadian shoppers should look for MK-7 listed on product labels to ensure the 120 microgram allowable dose will provide the health benefits they seek.

A final note on K₂-containing supplements: since menaquinone is a fat-soluble nutrient, look for a product that comes in a soft gelatin capsule or an oil-based, liquid suspension, instead of a hard capsule or tablet. This will provide the K₂ in a lipid- (fat-) based delivery system to enhance bioavailability. As with other fat-soluble nutrients, taking your K₂ supplement with food will also boost absorption.

Menaquinone-4 versus Menaquinone-7

	Menaquinone-4 (MK-4)	Menaquinone-7 (MK-7)
Source	Synthetic	Natural (natto)
Recommended dosage	45 milligrams (45,000 micrograms)	120 micrograms or more
Dosing frequency	Divided dose, three times daily	Once daily
Half-life in body	A few hours, hence the need for frequent doses	A few days, so a one-a-day dose is fine
Health Canada allowable dosage	Therapeutic dose of this form exceeds allowable dosage for Canada, suitable for United States only	Therapeutic dose within Health Canada limits of 120 micrograms

Vitamin K₂: Friend or Foe of Blood-Thinning Medications?

Oral anticoagulant (OAC) medications are prescribed to millions of North Americans to prevent heart attacks and strokes. Remember the vitamin K₁ cycle discussed in Chapter 2? Warfarin-(better known by the brand name Coumadin®, among others) type blood thinners block the body's recycling of vitamin K_1. This effectively creates a deficiency of vitamin K_1, so a portion of vitamin K_1–dependent blood clotting factors can't be activated. This, in turn, lowers the likelihood of forming a potential artery-blocking clot.

Patients on OAC therapy are instructed to avoid eating green leafy vegetables and a long list of other healthy foods that are high in vitamin K_1, since these foods could diminish the effect of the medications by providing K_1. Patients on blood thinners also have their blood clotting capacity (called the "INR value") monitored closely to ensure that dietary intake of vitamin K_1 hasn't altered the effectiveness of their meds. Diet can cause wide fluctuations in blood clotting ability for individuals on OAC therapy.

Since a main goal of taking blood thinners is to prevent cardiovascular disease events (heart attack and stroke), would it surprise you to learn that the long-term side effect of these medications is an unsafe buildup of plaque in your arteries and loss of bone density? Indeed, the side effects of warfarin treatment spills over to the domain of vitamin K_2, increasing the likelihood of atherosclerosis and osteoporosis in people who take these medications for prolonged periods. Just as blood thinners limit the activation of K_1-dependent clotting protein, they seem to hinder the activation of life-preserving MGP and osteocalcin.

Warfarin therapy and the Calcium Paradox

Warfarin therapy provides an accelerated model for developing the deadly Calcium Paradox. It illustrates exactly what happens to the body when (whether or not you are taking calcium supplements) you inhibit the most important nutrient to govern calcium deposition. This is another factor that makes vitamin K_2 absolutely unique. We can pinpoint the consequences of K_2 deficiency in a way that isn't possible with any other nutrient.

Given that the mechanism of warfarin is to block the action of vitamin K, it would seem a no-brainer that any type of vitamin K supplement (K_1 or K_2) should be avoided while on these meds. Well, here's the part that will make your haematologist flip his lid: patients on oral anticoagulant therapy who take up to 50 micrograms of MK-7 per day have more complete carboxylation of osteocalcin without interfering with the effect of the blood thinner. In other words, taking a small amount of MK-7 allows you to avoid the side effects of these meds without interfering with their intended benefits. In fact, unstable control of blood clotting capacity is linked to low vitamin K intake.[20] Taking a small amount (less than 50 micrograms) of menaquinone minimizes the fluctuations in clotting capacity induced by diet. More predictable blood clotting with fewer side effects is certainly a win-win scenario.

The race is on to find oral anticoagulants that work by a mechanism other than vitamin K inhibition, as there are so many challenges associated with this type of blood thinner. In the meantime, if you are on conventional blood thinners, consult your doctor before taking a K_2 supplement, since taking more than 50 micrograms of K_2 might interfere with your prescription. Make sure your doctor is up to date with the most current information about K vitamins before she balks at the idea of your taking vitamin K_2.

While the introduction of processed food into our diets triggered the decline in our vitamin K_2 intake, industrialization of the food supply over the last century propelled us into K₂ deficiency. The major manifestations of the Calcium Paradox, osteoporosis and heart disease, are the glaring legacy of this shift. Other common age-related concerns are also directly attributable to a lack of K₂. Chapter 4 explains how a deficiency of K₂ lies at the very heart of why we age and takes an in-depth look at the antiaging benefits of maximizing your K₂ intake.

Vitamin K₂: The Ultimate Antiaging Vitamin

The term "antiaging" is thrown around a lot these days. The quest for the fountain of youth seems to have reached a frenzy in recent years. We'd all like to add years to our life and life to our years, but very few nutrients make a significant contribution to achieving that goal. Here we'll look at how K_2 helps delay or even turn back the clock on such ravages of time as osteoporosis, heart disease, Alzheimer's, varicose veins and wrinkles. But first, a practical look at the aging process itself, and how an optimal K_2 intake is central to maximizing longevity.

The Triage Theory of Aging

Why do we eventually grow old and succumb to chronic disease? Many factors contribute to the aging process. Ultimately, an accumulation of damage to our genetic material (our DNA) and a decaying of our cells' energy production centers (the mitochondria) result in cell death and our physical decline. DNA injury happens throughout our lives, but the body has mechanisms to repair the damage as it occurs to ensure healthy ongoing cellular function. Why would DNA not be repaired as it is supposed to, leading to physical degeneration and aging? That is the subject of the triage theory of aging, the most current understanding of factors that contribute to our senescence, or biological aging.

The term "triage" means the assigning of degrees of urgency to determine the order of treatment. If you have ever visited the emergency room, you were probably seen by a triage nurse, who performed a quick evaluation of your condition. The nurse then decided whether your broken arm, say, would be treated before or after some other patient's abdominal pain. Triage effectively rations treatment when resources are insufficient for everyone's need to be met at the same time. Treatment of less urgent concerns are delayed, while the most pressing needs receive immediate attention.

The triage theory of aging refers to how the body deals with nutrient deficiencies. Specifically, it asserts that "when the availability of a micronutrient is inadequate, nature ensures that micronutrient-dependent functions required for short-term survival are protected at the expense of functions whose lack has only longer-term consequences, such as the diseases associated with aging."[1] In other words, when dietary vitamins and minerals are not plentiful, they are used preferentially for immediate health maintenance and reproduction. The trade-off is that DNA repair is disabled, leading to increased accumulation of DNA damage, degenerative disease and earlier death over the long term. In short, dietary deficiencies accelerate the aging process.

The triage theory of aging is the brainchild of Bruce Ames, Ph.D., professor of biochemistry and molecular biology at the University of California, Berkeley.[2] Dr. Ames observed that many kinds of vitamin and mineral deficiencies cause long-term DNA damage without noticeable short-term effects. How could that be? Why would Mother Nature allow us to slide along with inadequate nutrient intake and not alert us to the problem with signs or symptoms? Apparently, Mother Nature triages for survival. Nutrients are shunted toward immediate needs to increase the chance of short-term survival and reproduction. Longevity and long-term vitality are sacrificed, if necessary.

It's a nice theory, but does it hold true in real life? Is there any evidence that less than optimal levels of a given nutrient could go unnoticed while leading to long-term illness? It just so happens that the first nutrient researchers focused on to assess the validity of the triage theory of aging was vitamin K. The K family of vitamins lends itself to this analysis because, unlike other nutrients, K vitamins have only one function: to carboxylate (activate) vitamin K–dependent proteins. The health effects of this nutrient, or a lack thereof, are far-reaching because of the actions of these special proteins. Even so, it is relatively easy to assess whether our intake of K vitamins is sufficient to meet our health needs: just measure whether vitamin K–dependent proteins are fully activated or not. Specifically, in order to test the triage theory, we need to understand exactly what happens when we are slightly lacking in vitamin K.

When people are put on vitamin K–deficient diets in experimental situations, under-carboxylated osteocalcin levels increase weeks before under-carboxylated coagulation factors increase.[3] The body assigns more urgency to blood clotting, diverting limited vitamin K resources to that function, and bone health quietly suffers because of it. This illustrates the triage theory perfectly: you have to be desperately deficient in vitamin K₁ for blood clotting problems to manifest, whereas a slight lack of K₂ allows osteoporosis and atherosclerosis to progress silently for decades before symptoms appear.

Scientists now conclude that long-term vitamin K₂ inadequacy is "an independent, but modifiable risk factor for the development of degenerative diseases of ageing, including osteoporosis and atherosclerosis."[4] In other words, apart from any other contributing factor for chronic disease, not having enough vitamin K₂ throughout life to completely carboxylate *all* of your vitamin K₂–dependent proteins will increase your risk of illness down the road. The good news is that you can bump up your intake of vitamin K₂ and thereby lower your risk of age-related conditions such as bone fragility, arterial and kidney calcification, cardiovascular disease and cancer.[5]

The triage theory of aging could also be called the triage theory of nutrient requirements. This concept exposes the reality that we've been taking the wrong approach to setting recommended daily intake (RDI) levels. RDIs are based on the very minimum amount of a given nutrient that is needed to prevent an acute (short-term) deficiency. According to the triage theory, if we have less than optimal levels of any nutrient, even for a single day, we will pay the price in long-term injury. That explains why so many apparently healthy people can get by with vitamin K_2 deficiency—they're going to pay for it later. Let's take a look at the major age-related concerns of vitamin K_2 insufficiency, and the health benefits you can expect from maintaining an optimal menaquinone intake.

Vitamin K_2 for Heart Health

Did you breathe a sigh of relief after your last checkup because you were told you had normal cholesterol levels? Were you reassured that you aren't at risk for heart disease? Consider this: 50 percent of people who have heart attacks have normal cholesterol.[6] It turns out that chasing blood lipids has led us astray in the fight against heart disease, the leading cause of death in North America. Your cholesterol may be high or low, but what really matters is whether calcium-rich plaques are building up in your arteries, leading to potentially fatal blockages. Research now confirms that vitamin K_2 is the single most important nutritional factor in preventing *and even reversing* arterial blockages. Indeed, ensuring adequate vitamin K_2 (menaquinone) intake might be the most important thing you can do to extend your life. Before delving into the critical importance of K₂ for heart health, let's consider the paradigm shift around scientific understanding of the origins of cardiovascular disease.

Rethinking Cholesterol

Ask anyone why having high cholesterol is bad and he or she will probably tell you, "Because it causes heart disease." With cholesterol-lowering medications being the number one prescribed drug, many doctors have treated cholesterol itself as the disease and seemingly lost sight of the fact that cholesterol doesn't actually *cause* heart disease—it never has. High cholesterol is only one risk factor associated with developing the arterial blockages that can lead to heart attack and stroke. Somewhere along the way, correlation and causation got crossed up, and cholesterol was found guilty, without due process, of causing heart attacks and strokes. (For a fascinating look at the history and politics of how the erroneous sat-fat heart health theory took hold, I heartily recommend the book *Good Calories, Bad Calories*, by Gary Taubes.)

Blaming cholesterol for heart disease is a lot like blaming firefighters for a fire. Although the increased levels of some kind of blood lipids (fats) are imprecisely associated with a greater chance of heart disease, that doesn't mean they caused the problem. In addition to its many other essential roles in the body—like making vitamin D from sunshine—cholesterol is an anti-inflammatory compound. More than anything, high intake of sugar and refined flour will cause inflammation in the body, causing cholesterol levels to rise in an attempt to sooth the inflammation. If, over time, increasing cholesterol can't compensate for a poor diet, heart disease sets in and poor old cholesterol gets the blame. But cholesterol didn't cause the problem—it was trying to help! Cholesterol's real contribution to heart disease has been blown way out of proportion, and many experts seem to have blinders on when it comes to this particular nutritional compound.

Contrary to how we tend to speak about cholesterol, there aren't really different types ("good" and "bad"). There is only one type of cholesterol, but it travels in different vehicles in the blood, just as you are only one person but can travel in different cars. High-density lipoprotein (HDL) is the vehicle that takes cholesterol from the body's tissues back home to the liver, like an efficient, nonpolluting, electric car. Having high amounts of HDL is a good thing; it protects against heart disease. Low-density lipoprotein (LDL) is the other main carrier, but it takes cholesterol from the liver out to the rest of the body. Although LDL has been called "bad" cholesterol, it is really only when LDL is oxidized (damaged) that it becomes a problem. Having lots of oxidized LDL is like driving around in a rusty-old beater that is spewing fumes and dropping pieces as it goes: not good for the environment.

Blood lipid levels can be a helpful indicator of the risk of heart disease, if they are used appropriately. Total cholesterol (combined HDL and LDL) is not useful. Is it bad to own 100 cars? That depends in part on whether they are all efficient hybrids or rusty beaters. Having high (over 1.0) HDL is good—the more the better when it comes to HDL. If you are told you have high (over 2.6) LDL, ask for a lipoprotein(a) test, which will tell you if that LDL is oxidized or not.

Cholesterol should certainly not be ignored altogether. When LDL levels are sky-high, it is a sign that the body is being triggered to produce too much cholesterol or it is not using cholesterol efficiently, or both. This does need to be addressed, but not by merely cutting off the body's production of cholesterol. Lowering cholesterol by taking a statin medication, a drug that halts the liver's cholesterol synthesis, is shortsighted, like sending the firefighters home while the fire blazes on. By addressing the factors that prompted the rise in cholesterol, such as insulin resistance, inflammation and oxidative stress, we can target the underlying causes of abnormal cholesterol. In other words, when the firefighters show up, we need to take steps to put out the fire.

Having normal cholesterol isn't the end of the story, either. Focusing on cholesterol to the exclusion of other risk factors will result in missing 50 percent of silent heart disease. After all, we need to keep in mind that even when cholesterol is perfectly normal, calcium-containing arterial plaque could be quietly narrowing the blood vessels that feed the heart.

Atherosclerosis: Plaque Accumulation

The word "atherosclerosis" comes from the Greek words *athero* (gruel or paste) and *sclerosis* (hardness). It refers to the deposit of calcium, fatty substances and scar tissue (collectively known as plaque) that builds up in the inner lining of an artery. A substantial buildup of plaque can significantly reduce blood flow through an artery. Alternatively, arteries can become fragile and rupture, leading to blood clots that can block blood flow or break away and travel to another part of the body. If the clot blocks a blood vessel leading to the heart, a heart attack results. If the clot blocks a blood vessel leading to the brain, the result is a stroke.

Atherosclerosis leads to coronary artery disease, also known as heart disease. Coronary artery disease is the result of plaque accumulation within the walls of the coronary arteries, the arteries that supply oxygen and nutrients to the muscles of the heart. Coronary artery calcification is part of the development of atherosclerosis and refers to the appearance of calcium deposits in the atherosclerotic plaques by a process that mirrors bone formation. Calcium plaque can develop as early as the second decade of life, but it occurs with more frequency in older age. The presence of calcium plaque is a much better predictor of heart attack risk than cholesterol, and tests are now available to measure it.

How spreadable is your butter?

"Probably every housewife is familiar with the low melting quality of the butter produced in early summer when the cows have been put on green pasture. This is particularly true of butter that has the grassy flavor and the deep yellow to orange color."[7] So said Dr. Weston A. Price in 1939.

Soft, orange butter? It's hard to believe that just two generations ago, familiarity with seasonal variations in butter quality was common knowledge. Nowadays, cold butter is pale, firm and difficult to spread year-round, while soft butter is even harder to come by than housewives.

Does butter straight from your fridge spread evenly or does it clump up and tear the bread? The spreadability of your butter is directly related to the cow's diet. Soft butter doesn't transport well, so, in the 1930s, dairy producers added cottonseed meal and other grains to the cows' diet to make firmer butter that shipped more easily. When the dairy industry switched over to primarily feeding corn, soy and cereals in the 1960s, the softness "problem" took care of itself, and butter that spreads easily straight from the fridge became a thing of the past.

Now it takes scientific studies to relearn what was once common sense: grass feeding makes better butter. According to recent research, fresh grass in the cow diet improves the texture, flavor and nutritional properties of butter.[8] Specifically, the more grass a cow has in her diet, the lower the atherogenicity index of the milk she produces. This index ranks the nutritional elements in food that are associated with heart disease. Soft, spreadable butter from cows that eat grass is lower in saturated fat, higher in unsaturated fatty acids and higher in vitamin K_2 than conventional butter. This butter is not just part of a heart-healthy diet; it could very well be the foundation. The effect of grass feeding on the quality of dairy products points to a very important reality: labeling butter, lard and egg yolks as "artery-clogging" fare is more a reflection of how those foods are produced than of their essential nature. Once again, we are not just what we eat but also what our animals eat.

Regardless of the results of your cholesterol screening, there is one crucial question you really need to ask your doctor: "Do I have calcium plaque in my arteries?" The answer to this question used to involve a lot of indirect guesstimates and a wait-and-see approach. Now you can know for sure. A new test is rapidly replacing cholesterol levels as the gold standard for assessing heart disease risk. The coronary artery calcium score, also called cardiac calcium score or just calcium score, is a special type of X-ray that checks for the buildup of calcium in the coronary arteries; Chapter 6 tells you what you need to know.

What Really Causes Heart Disease?

So, if cholesterol is not the best predictor of heart disease and arterial calcium buildup is the real problem, what causes arterial calcification? The research is loud and clear: a lack of vitamin K_2 is the single biggest dietary factor for cardiovascular disease risk. Multiple lines of evidence converge on this nutrient.

Population-based studies draw a definitive connection between low K_2 and risk of heart disease. For example, the Rotterdam Study followed 4,600 men aged 55 and older in the Netherlands. Men with the highest intake of K_2 had a 52 percent lower risk of severe aortic calcification, a 41 percent lower risk of coronary artery disease, a 51 percent lower risk of dying of coronary artery disease and a 26 percent reduced risk of total mortality. Researchers determined that an adequate intake of vitamin K_2 could be extremely important for prevention of coronary heart disease. Two more studies published in 2009 echoed these findings: researchers determined that adequate K_2 intake reduces both coronary artery calcification and the risk of coronary heart disease.[9]

Laboratory research also shows that vitamin K_2 is the strongest inhibitor of tissue calcification that we know of, and if you aren't getting enough, your heart health will suffer. In Chapter 1's discussion of how vitamin K_2 works, I mentioned the process of carboxylation. A simplified definition of this complicated-sounding process is that carboxylation activates proteins so they can work. Vitamin K_2 activates many important proteins, including one called matrix gla protein (MGP), which prevents calcium from depositing on blood vessels and other soft tissues. Healthy, plaque-free arteries contain plenty of K_2-activated MGP. Diseased, calcium-laden, atherosclerotic blood vessels are full of inactive MGP, because K_2 wasn't there to turn it on. Vitamin K_2 is such a crucial determinant of heart health that inactive MGP is directly correlated with the severity of coronary artery calcification. In other words, the more you are lacking vitamin K_2, the less MGP will be activated to bind calcium, so calcium just lodges in arteries, creating blockages. Since only vitamin K_2 activates MGP, taking it regularly is one of the most important preventative measures you can follow to ward off heart disease.

Activating MGP isn't the only way K_2 prevents heart disease. Several other vitamin K_2–dependent proteins also play a role in protecting against atherosclerosis. Growth arrest-specific gene 6 (Gas6), for instance, promotes the rapid clearance of dead smooth muscle cells that can act as an anchor for circulating fats within arteries. In addition, vitamin K_2–dependent protein S encourages the immune system to gently take out the arterial garbage rather than launch a full-on inflammatory attack, which can encourage plaque formation. It isn't necessary to get bogged down in the complex details of how menaquinone keeps your blood vessels clear and healthy; just be aware that vitamin K_2 prevents cardiovascular disease in many ways.

Vitamin K_2 Trumps K_1 for Heart Health

The same population-based studies that show vitamin K_2 is crucial for preventing heart attacks also make it clear that the intake of vitamin K_1 (phylloquinone) is not related to heart health.[10] Why is that significant? Because it shows that, although eating fruits and veggies may lower your risk of cardiovascular disease by providing fiber and antioxidants, you can't rely on the type of vitamin K found in these foods to prevent atherosclerosis. The best food sources of heart-healthy menaquinone are natto and egg yolks, butter and fat from grass-fed animals. But wait, aren't those the things we were told to avoid for years to minimize the intake of "artery-clogging" saturated fat? Sadly, yes, and that advice was just plain wrong. There was never much compelling evidence to link saturated fat to heart disease—although the notion did help to sell a lot of margarine. A 2010 meta-analysis published in the *American Journal of Clinical Nutrition* concluded that there is insufficient evidence connecting saturated fat to coronary heart disease.[11] Ironically, jumping on the vegetable oil bandwagon did us no good in the fight against heart disease, and it further depleted our K_2 status. Had we instead stuck with the old-fashioned ways of farming and eating, we would have been better off.

You *Can* Reverse Heart Disease

Lots of lifestyle changes can help prevent heart disease (lose weight, cut your sugar intake, exercise), but can any substance remove calcium plaque once it has formed? Just one: vitamin K_2. Studies show that adding menaquinone to the diet will activate MGP to reduce arterial calcium content by 50 percent over just a six-week period. This cardiovascular news just keeps getting better, since the same studies show that blood vessels are not irreparably damaged by the plaque, as you might expect. Apparently, vitamin K_2 also helps restore arterial flexibility once the calcium has been removed.[12] If you have a high coronary artery calcium score or elevated levels of inactive osteocalcin, take heart: vitamin K_2 can help.

The coronary arteries aren't the only blood vessels in and around the heart that succumb to perilous calcification. Very seriously, plaque can build up in the aorta, the major blood vessel that carries fresh, oxygenated blood from the heart out to the body. This causes the aorta to become rigid and inflexible, increasing the risk for heart attack. Aortic stiffness also precedes kidney disease, an equally grave condition that is covered in the next chapter. Vitamin K$_2$ is just as effective at removing calcium from the aorta as from the coronary arteries, as illustrated in the case of Sam K., a 69-year-old dentist with a heart murmur.

Sam's primary care physician detected the abnormal heart sound during a routine physical examination. He therefore had Sam undergo an echocardiogram, a simple test useful to evaluate disorders of the heart valves. The echocardiogram showed that Sam had aortic valve stenosis, a condition in which calcium and other material deposited on the aortic valve cause it to stiffen. A stiff aortic valve struggles to open with each heartbeat and can obstruct the blood output of the heart. This leads to chest pain, breathlessness, lightheadedness and heart failure. Although symptoms at first occur with vigorous physical activity, as the valve gets stiffer, symptoms occur with minimal physical provocation. The severity of aortic valve stenosis is gauged by measuring the effective area of the valve opening. Normal is 3.0 centimeters squared; Sam's aortic valve area was reduced to 1.6 centimeters squared, about half of what is should have been.

Aortic valve stenosis is eventually fatal. For this reason, once it's identified, an echocardiogram is repeated every 6 to 12 months. When the valve opening is reduced to 1.0 centimeters squared or less and symptoms begin, aortic valve replacement is advised. This is an open-heart surgery, a major undertaking at any age. Because most people with aortic valve stenosis are in their 70s and 80s, an open-heart procedure carries substantial risk. Efforts have been made over the years to identify treatments that slow the progression of aortic valve disease.

The only agent that has shown any effect in slowing aortic valve stenosis is high-dose Crestor, a potent cholesterol drug. The dose used in the study, 40 milligrams per day, carries crippling side effects for most people.

Sam had the good fortune of being referred to a forward-thinking cardiologist, Dr. William Davis, who, since 2006, had been advising patients to supplement vitamin D to prevent progression of aortic valve disease.[13] Achieving a therapeutic blood level of vitamin D meant a dose of 8,000 international units per day for this average-sized man. The specialist found that high-dose vitamin D alone stopped the aortic valve area from shrinking in over 90 percent of his patients, although it did not reverse the existing disease.

Sam is a nutritional supplement enthusiast, so when his doctor told him about the benefits of vitamin D to aortic valve disease, he jumped on the idea. At the time, Sam's cardiologist also suspected that vitamin K_2 supplementation would add an additional advantage. Among the observations that pointed toward vitamin K_2 as a factor in aortic valve disease was that people who take the blood-thinning drug warfarin, or Coumadin—which induces both vitamin K_1 (associated with blood thinning) and K_2 (associated with calcium metabolism) deficiencies— experience gradual calcification and narrowing (stenosis) of their aortic valves. Because he loved the idea of applying nutritional supplements in a rational, targeted way, Sam added to his vitamin D supplementation 900 micrograms of the short-acting MK-4 form of K_2 and 100 micrograms of the long-acting MK-7 form, along with 1,000 micrograms of K_1, to cover all his vitamin K bases. In reality, the dose of MK-4 was not likely therapeutic and the K_1 wasn't really necessary, but the 100 micrograms of MK-7 provided effective treatment.

Ten months later, another echocardiogram showed an aortic valve area of 2.9 centimeters squared—nearly doubling the valve area. The finding was so remarkable that Sam's doctor asked the echocardiography technician to confirm precisely what he had found. Yes indeed, by using a combination of vitamins D and K, Sam had managed to open up his valve to essentially a normal, healthy size.

How Much Is Enough?

Sam's story illustrates an example, from the very early days of menaquinone awareness, of using K_2 to treat an advanced form of heart disease. Most of us are more concerned with preventing catastrophic calcium buildup. But how do you know if you have sufficient vitamin K_2 to ensure all of your MGP is actively preventing heart disease? Although there are blood tests to measure levels of inactive MGP, they are used almost exclusively in research settings and are not readily available to patients and practitioners. Fortunately, rapid, reliable screening tests are just around the corner. In fact, emerging evidence suggests that this simple blood test is such a good indicator of calcium buildup in the arteries that it will likely replace the coronary artery calcium score, saving you the minor hassle of having to have a CT scan of your heart. In the meantime, since vitamin K_2 has no known toxic effect, load up on K_2-rich foods and include a K_2 supplement in your diet.

Even More Evidence for Vitamin K_2 and Heart Health

Dr. Weston Price knew, over 70 years ago, that a lack of vitamin K_2 caused heart disease. He measured samples of dietary activator X (now known to be K_2) from multiple regions of North America throughout the year.

He showed that K$_2$ levels increased during the summer (when cows were eating more grass), while rates of death from heart attack decreased. That trend reversed in the winter: as K$_2$ intake naturally declined in the winter months, death from heart attacks increased. Price conducted similar studies using samples from Australia, New Zealand and other countries in the southern hemisphere and saw the very same trend in parallel with the seasons.[14]

Remarkably, 2010 studies confirm this seasonal variation in blood vessel calcification. In the northern hemisphere, calcified plaque on coronary arteries is highest in January and February, and lowest in August.[15] Seasonal change accounts for up to 3 percent of changes in blood vessel calcification. This might not sound like a lot, but it mirrors an annual cycle in cardiac mortality, the same cycle charted by Price. This points to the critical role of fat-soluble vitamins—whose dietary levels vary naturally with the length of daylight hours—in heart disease prevention. Fortunately, we now have the advantage of vitamin supplements, so we are not completely at the mercy of seasonal cycles to dictate our intake of fat-soluble nutrients.

Although modern research confirms the yearly cycle of cardiovascular disease, Price's work is unique because he compares this effect in a dozen different regions. If cardiac death were only a matter of vitamin D levels, it would show a more or less identical pattern in every region of the northern hemisphere. No matter where you live in that hemisphere, days are longest in June and shortest in December, and unsupplemented vitamin D levels will vary accordingly. Length of daylight hours isn't the chief controlling factor, however; something else is involved: cardiac mortality roughly follows annual vitamin D levels but precisely follows vitamin K$_2$ levels.

Vitamin D Deficiency: Another Risk Factor for Cardiovascular Disease

Of all the things you get tested for during your annual checkup to supposedly monitor heart disease risk (HDL, LDL, triglycerides, body weight, etc.), I'll bet that your vitamin D level is not one of them. It should be. Although vitamin D might not be the principal nutrient governing heart disease, people who are deficient in vitamin D have a significantly increased risk of heart attack compared with those with sufficient vitamin D levels.[16] A lack of vitamin D seems to influence cardiovascular disease in several ways, and a key mechanism is its close relationship with K₂. Vitamin D is required to make MGP, the most important protein to prevent artery calcification. Vitamin K₂ activates the MGP made by vitamin D, so these nutrients team up to keep arteries free of calcium plaque.

Vitamin K₂: Building Better Bones

Are you among the millions who have been dutifully gobbling up calcium supplements for years, only to be told that your bone density is still too low? Osteoporosis afflicts more than half of North Americans aged 50 and older. This disease is responsible for millions of fractures annually, primarily involving the spine, wrists, ribs and, most seriously, hips. In a person with osteoporosis, a hip fracture often occurs during a relatively minor fall. This can take months to heal and involve many complications, even death. Despite a drastic increase in our calcium supplementation and consumption of calcium-fortified foods of all kinds, rates of osteoporosis and hip fracture remain unchanged. Why? Because taking just calcium alone doesn't guarantee it will end up in the right place. We also need vitamin K₂, which helps get the calcium into our bones.

What about vitamin D? Supplementing with both vitamin D and calcium has been shown to reduce hip fracture risk somewhat. However, for many people, these supplements still don't make enough of an improvement in bone density. Once again, vitamin K_2 is the missing link. Studies show that taking vitamin K_2 and vitamin D together improves bone density and reduces fracture risk more than either nutrient alone. In fact, most of the widely publicized benefits of vitamin D on our bones are really dependent on vitamin K_2. If you have been taking calcium and vitamin D without vitamin K_2, you—and your bones—are missing out.

The Renovation of You

Although we think of our skeleton as solid and unchanging, bone tissue is constantly being modified and maintained by a process called remodeling. This happens thanks to the actions of cells known as osteoblasts and osteoclasts. Osteoblasts are the bone-building cells (just remember, the "b" in osteoblast stands for "build"). Osteoblasts originate in bone marrow, and play a role in mineralization (calcium deposition) of the bone matrix, the structural storage area that holds calcium within bone. Osteoblasts are crucial not only for bone building but also for maintaining and strengthening existing bone. Osteoclasts, on the other hand, are cells that help to tear down bone. The word "osteoclast" comes from the Greek *osteon* (bone) and *klastos* (broken). The breakdown process is known as bone resorption, and while osteoclasts might sound like nasty cells, they do help to remove damaged bone. Together with osteoblasts, these cells eliminate areas of weakness and repair cracks and fractures to keep bones strong and healthy. Maintaining a healthy balance between the action of osteoclasts and osteoblasts is essential to preventing osteoporosis.

The balance between resorption and deposition of calcium into bone changes with age. In growing children, bone formation exceeds breakdown, so bones get longer and stronger. In our young adulthood and midlife years, the processes are roughly equal, to maintain bone density. As we get older, however, and due to factors like genetics, not enough exercise or a lack of key dietary nutrients, bone breakdown can exceed formation, resulting in a loss of bone mineral density—either osteopenia (mild bone loss) or osteoporosis (moderate to severe bone loss).

Calcium Alone Is Not Enough

Since bones are primarily composed of calcium, a standard recommendation for those with osteoporosis is to take calcium supplements. When that doesn't work, patients are often told to take more and more calcium. Currently, most doctors are suggesting a whopping 1,500 milligrams of calcium per day for their patients with low bone density. However, taking calcium supplements by no means ensures that it will end up where it is needed. Indeed, as I mentioned earlier, studies show that calcium supplements may very well be causing plaque in coronary arteries. So how can you be sure that the calcium you take is really helping your bones and not harming your heart? By also taking calcium's best bone-building buddy, vitamin K₂, along with vitamins D and A.

Fat-soluble vitamins play critical roles in developing strong bones. Vitamin D₃ (cholecalciferol) is made in your skin from cholesterol when it is exposed to ultraviolet B light from sunshine. A key role of vitamin D is to increase the body's absorption of calcium. When there is insufficient vitamin D, the body can't absorb enough calcium, and bone density suffers. Vitamin D, along with vitamin A, is also necessary for the production of osteocalcin, a bone-mineralizing protein discussed in more detail below. It is important to note that, although vitamin D is crucial for calcium absorption, it has no effect on what happens to calcium once it is absorbed. That role belongs to K_2.

Deficiency of vitamin D is common. Our skin makes vitamin D_3 through a combination of cholesterol and ultraviolet B rays from the sun, but these are the same rays that are responsible for sunburn. Ultraviolet B rays are at their vitamin D–making best during midday hours at higher latitudes. This means that the best time for making vitamin D_3 is between 10 a.m. and 2 p.m. during summer months— which is exactly when skin cancer experts have told us to stay in the shade—or use heavy-duty sunblock. These "sun safety" recommendations, in addition to an indoor lifestyle, have played a major role in vitamin D deficiency. Fortunately, vitamin D awareness has increased dramatically in recent years, and most health experts now recommend taking between 1,000 and 5,000 international units of the vitamin daily.

A caveat about vitamin D

Despite the concerns about vitamin D deficiency, it is critical to be aware that too much vitamin D on its own can demineralize your bones. Vitamin D increases both the demand for vitamin K_2 and the potential to benefit from K_2-dependent proteins like MGP and osteocalcin.[17] To prevent vitamin D toxicity, always take vitamin D along with vitamins A and K_2. Many health experts still believe that vitamin A is actually harmful to bones but, again, that outdated notion is based on old studies where high amounts of vitamin A were given without vitamin D or vitamin K_2. Chapter 7 explores specific recommendations for how to balance all of the fat-soluble vitamins for maximum benefits.

Vitamin K_2: The Missing Link for Bone Health

While we've known for eons about the importance of calcium in bone health, and the function of vitamin D is now better understood, practitioners and patients have still struggled with osteoporosis. We finally have the last piece of the puzzle. Vitamin K_2's role in building bones truly makes it the unsung hero in maintaining bone health.

Just as with heart health, K_2's prime role in bone health is to carboxylate (activate) certain proteins, allowing them to bind calcium. There are several vitamin K_2–dependent proteins in bone, of which osteocalcin is the most abundant and best known. It is the main protein involved in the deposition of calcium into bones and teeth. Only once it is activated by vitamin K_2 can osteocalcin grab on to calcium and lay it into the bone matrix. Without sufficient vitamin K_2 to activate it, osteocalcin remains useless, calcium will not be deposited into the bone and osteoporosis sets in. That is why most of the bone-building benefits of vitamin D are really dependent on K_2: vitamin D, assisted by vitamin A, stimulates the production of osteocalcin, and K_2 activates it. Once again, fat-soluble nutrients collaborate to achieve optimal health.

It's obvious, then, how activated osteocalcin prevents brittle bones, but vitamin K_2 is a boon to bones in other ways as well. Research shows that K_2 partners with vitamin D_3 to inhibit the production of osteoclast cells that break down bone.[18] K_2 also targets osteoclasts to undergo apoptosis (programmed cell death), leading to a reduction in the number of osteoclast cells.[19] Impeding the action of osteoclasts in this way helps bone-building osteoblasts catch up to maintain that healthy balance.

Bone Health During Menopause

Osteoporosis becomes a major issue for women after menopause, when less estrogen is produced in the body. Estrogen impacts calcium metabolism in several ways. For example, it promotes the conversion of vitamin D to its active, bone-building form, so diminishing estrogen levels interfere with vitamin D activity. When estrogen levels drop, osteoclast activity increases. Even worse, the decline in estrogen causes an increase in a compound called interleukin-6, which stimulates the production of more osteoclasts.[20] Does this mean that women are doomed to suffer from crumbling bones as soon as they experience menopause? Fortunately, vitamin K₂ counteracts each of these problems: taking menaquinone-7 (MK-7) has been shown to compensate for the changes in bone density that are caused by menopause.[21]

Bone Health and Natto

"Drink milk for strong bones." That advice has been drilled into our brains since grade school. Yet the intake of dairy products and other high-calcium foods is much lower in Japan than it is in North America, and the Japanese don't suffer nearly the same rates of osteoporosis as we do here. That's because calcium intake isn't the most important factor in determining bone health. What do the Japanese do that we don't? They eat natto, the smelly superfood that's an abundant source of vitamin K₂.

Natto isn't widely eaten everywhere in Japan; it's more popular in the eastern part of the country. Studies show a statistically significant inverse correlation between the incidence of hip fractures in Japanese women and natto noshing in each prefecture throughout Japan. In other words, Japanese women who ate more natto had fewer fractures.[22] Further studies show that men with occasional or frequent dietary intake of natto have significantly higher blood levels of K_2 and activated osteocalcin concentrations than those who never eat the stuff. If all this good news doesn't make it any easier to stomach fermented beans at breakfast, you'll be glad to know that taking K_2 supplements has the same benefits.[23]

Bone Health and Organ Transplant

Osteoporosis is a major problem after organ transplants. A large and rapid decrease in bone mineral density occurs within the first year following almost all types of organ transplant. Organ recipients have up to 34 times higher risk of bone fracture than age-matched individuals who do not undergo transplants.[24] The problem involves several factors: the disease process or diseased organ itself might promote bone loss, and many common medications given before or after the transplant further decrease bone strength. That being said, no pretransplant factor predicts post-transplant fracture risk, and patients may sustain a fracture despite normal pretransplant bone mineral density. So, organ recipients just need to pile that onto their list of health concerns at an already vulnerable time. Or do they?

In 2010, Norwegian researchers reported that vitamin K_2 has a favorable action on bone health for transplant patients. The study included 35 lung and 59 heart recipients. The participants were randomly assigned to receive either a 180 microgram MK-7 supplement or a menaquinone-free placebo for one year after surgery. At the end of the study, bone mineral density in the lumbar spine and bone mineral content in the treatment group actually went up compared with the group that did not take menaquinone.[25] Remarkably, trial participants experienced the benefit even though many were deficient in vitamin D.

Since patients often wait two or more years before transplantation, there is an opportunity to prevent further bone loss and to help restore what may already have been lost.

Vitamin K_2 versus K_1 for Bone Health

Misleading information about vitamin K is everywhere, including magazine articles that tout green leafy veggies as a "great source for bone-building vitamin K." Yes, green leafy vegetables are a source of vitamin K, but not the kind that most efficiently prevents osteoporosis. Research shows that it takes 1,000 micrograms of a highly absorbable pharmacological preparation of K_1 daily to activate osteocalcin. Unfortunately, humans are incapable of absorbing even one-fifth of that amount from food. On the other hand, we are able to absorb large amounts of vitamin K_2 from foods. Studies done in the Netherlands indicate that vitamin K_2 was three times more effective than vitamin K_1 at raising activated osteocalcin numbers over a 40-day period. Most interesting was that the impact of vitamin K_1 leveled off after only three days, while the effect of K_2 *increased* every day of the study.[26]

Testing for Your Vitamin K_2 Levels

When there isn't enough K_2 in your blood to satisfy the K_2-dependent proteins, the inactive proteins circulate in the blood stream until they are destroyed. A simple blood test measures the amount of inactive osteocalcin, making it possible to determine whether your bone cells have sufficient menaquinone to meet their needs. Research shows that the risk of hip fractures is five times greater in people with the highest percentages of inactive osteocalcin. A bone density scan will tell you whether you have osteoporosis. An under-carboxylated osteocalcin test will tell you if you should be taking more K$_2$ to treat it. (Chapter 6 provides more information on how to test your body's vitamin K$_2$ levels.)

Multiple studies since the 1980s show a link between vitamin K$_2$ and bone health, and a recent study of nearly 900 adults found that those with the lowest levels of vitamin K$_2$ had a 65 percent greater risk of hip fractures compared with study participants with the highest K$_2$ levels. Although K$_2$ might be the most important and neglected nutrient for bone health, other vitamins and minerals are necessary to optimize the body's use of calcium. Vitamin D, vitamin A, magnesium, phosphorous, magnesium, zinc, boron and B vitamins all cooperate to enhance bone density and lower fracture risk.

Vitamin K$_2$ for Alzheimer's Disease

Maintaining a sharp mind tops the list of age-related concerns for most people. After all, what's the point of living longer if you can't engage with your loved ones or if you become a burden to them? Vitamin K$_2$ deficiency is also linked to Alzheimer's disease, the brain disorder on everyone's mind, except perhaps those who have it. Alzheimer's is a degenerative brain disease that is now the most common cause of dementia, a gradual loss of memory, judgment and functional ability. By 2050, the prevalence of Alzheimer's is expected to quadruple to 1 in every 85 people, over 40 percent of whom will require high-level care, with a heavy strain on caregivers. Experts estimate that "if interventions could delay both disease onset and progression by a modest 1 year, there would be nearly 9.2 million fewer cases of disease in 2050 with nearly all the decline attributable to decreases in persons needing a high level of care."[27]

Alzheimer's type senile dementia is closely associated with a well-established vitamin K_2–deficiency condition, osteoporosis. Alzheimer's is so strongly linked to accelerated bone loss that bone mineral density in early stages of the disease diminishes in tandem with gradually shrinking regions of the brain.[28] The researchers who discovered this connection suggested that the brain's degeneration might somehow disrupt the central mechanism of bone remodeling. An equally likely possibility is that a common factor is responsible for both bone loss and brain loss. That seems to be the case for the other face of the Calcium Paradox. Heart disease is also so strongly linked to Alzheimer's that the former can be considered the forerunner of the latter. The connection long evaded researchers, since many patients die of coronary artery disease before cognitive changes are apparent.[29]

People with Alzheimer's disease consume less than half the dietary vitamin K of their mentally healthy counterparts.[30] Alzheimer's patients have lower dietary intakes of vitamin K, lower bone density, more hip fractures and more severe vitamin K_2 deficiency than people of the same age without cognitive decline.[31] That may be suggestive, but it might just reflect generally poorer nutrition in Alzheimer's patients. What evidence suggests that vitamin K_2 deficiency is directly related to Alzheimer's, or that supplementing with K_2 might help prevent or treat it? To answer this question we have to look more closely at the known factors contributing to the disease.

Exactly what causes the most common form of dementia, and what causes it to progress, isn't well understood. On autopsy, the brains of Alzheimer's patients contain characteristic "plaques and tangles," which are different types of abnormal protein deposits within the brain tissue. Exactly how the plaques get there is unclear, but at least two factors contributing to their presence pertain to vitamin K_2: free radical damage and insulin sensitivity in the brain. Multiple lines of evidence suggest that oxidative stress is significant in the onset of Alzheimer's and development of those characteristic brain lesions.[32] Although vitamin K_2 isn't known to act as an antioxidant anywhere else in the body, it has a powerful ability to prevent free radical formation within the brain.

Free radicals are highly unstable, reactive molecules that rob our tissues and cells of electrons in order to achieve stability. In doing so, they cause oxidative damage to our cells in a process that is very similar to metal rusting. An accumulation of free radical damage is the underlying mechanism of many degenerative diseases and is central to the age-associated DNA damage mentioned earlier.

Normally, protecting against free radical damage is the job of antioxidants. Antioxidants are molecules, often dietary nutrients, that scavenge renegade radicals roaming the body and causing destruction at a cellular level. Antioxidants accomplish this by generously donating electrons to stabilize free radicals. Vitamin K_2 is not a classical antioxidant, meaning it doesn't donate electrons, yet somehow it completely blocks free radical accumulation and brain cell death in laboratory studies.[33] Instead of just mopping up free radicals in the brain, vitamin K_2 prevents them from being made in the first place.[34] It also protects brain cells from a depletion of glutathione, a major antioxidant in the body.

Curious about which of our main vitamin K_2–dependent proteins, osteocalcin or MGP, is responsible for this remarkable health benefit? Neither. Vitamin K_2 completely blocks free radical accumulation and cell death by a mechanism that is independent of its only known action of gamma-carboxylation. This is excellent news for stroke victims or TIA (transient ischemic attack or recurrent mini-stroke) sufferers who have been put on blood thinners. If K_2 protected against brain damage only by its usual protein-dependent mechanism, then oral anticoagulant meds would interfere with that benefit, but they don't. Vitamin K_2 shields brain cells from free radical damage, even in the presence of the blood thinner warfarin.[35]

Vitamin K deficiency is also linked to Alzheimer's in another intriguing way: the production of and sensitivity to insulin in the brain. Unlike cells in the rest of the body, brain cells do not require insulin to absorb blood sugar. Sugar is so critical for brain cell function that it enters neurons without the need of insulin to unlock the door as it does in the rest of the body. For this reason, scientists long believed there was no connection between insulin and the brain—that the brain was insensitive to insulin. Now we know that isn't true. Insulin is very important for brain function and instrumental in learning and forming memories.

The Alzheimer's brain is much like the diabetic body. Either the brain isn't producing enough insulin, as with type 1 diabetes, or the brain cells become insulin-resistant, as with type 2 diabetes. The brains of Alzheimer's patients don't use glucose properly; indeed, the disease is now being called type 3 diabetes. Administration of insulin significantly improves the cognitive performance of Alzheimer's patients, and it stands to reason that improving insulin sensitivity and production in the brain will help prevent, delay or even reverse Alzheimer's symptoms.[36]

Vitamin K_2 deficiency increases our risk for diabetes, as you'll see in Chapter 5, and that alone may predispose us to Alzheimer's. Diabetics have a 30 percent to 65 percent higher risk of developing Alzheimer's disease than nondiabetic people. It's likely that there is an even more direct connection between this form of dementia and vitamin K_2. Menaquinone may increase the brain's production of and sensitivity to insulin, as it does elsewhere in the body, and it may prevent the generation of free radicals that contribute to senile plaques and tangles. The most current medical hypotheses suggest that vitamin K_2 deficiency contributes directly to the origination and development of Alzheimer's and that menaquinone supplementation will be beneficial in preventing or treating the disease.[37] If K_2 slows the development of Alzheimer's disease by even that one modest year, it could make a major difference to the well-being of our entire society.

Vitamin K₂ for Wrinkle Prevention

Fine lines, laugh lines, crow's feet—whatever you call 'em, we don't like 'em. And what's more, they are far from being just a benign, inevitable fact of aging. A growing body of evidence links the degree of skin wrinkling to the severity of health conditions such as osteoporosis, heart disease, diabetes and poor kidney function—diseases that are also all linked to a lack of vitamin K_2. For example, cutting-edge research shows that the degree of facial wrinkling provides a glimpse into bone mineral density. Specifically, the severity of a postmenopausal woman's facial wrinkles predicts her risk of osteoporosis. Women with extensive facial wrinkles are much more likely than their peers to suffer from low bone mass, while those with firmer skin tend to have denser bones, regardless of age or body weight.[38]

In a similar example, Korean research published in the journal *Nephrology* in 2008 found that increased facial wrinkling is associated with a reduced kidney filtration rate (a measure of kidney function), independent of age and sex.[39] American research published the following year demonstrated that decreased kidney filtration predicts an increase in inactive MGP—in other words, vitamin K_2 deficiency.[40] When it comes to skin, it seems that a K_2 deficiency might be written all over your face.

Epidemiological evidence shows that Japanese women have fewer wrinkles and less skin sagging than North American women of the same age. Given the vast difference in diet and lifestyle, perhaps this isn't a fair comparison. However, even among Asian cities, female residents of Tokyo have the least visible signs of aging compared with their age-matched counterparts living in Shanghai and Bangkok.[41] Granted, many Japanese ladies religiously avoid sun exposure, carrying parasols or wearing facial visors while outdoors. That being said, the inter-Asian groups are comparable for many other diet and lifestyle factors save one: the consumption of natto. Tokyoites commonly enjoy the pungent, fermented soybeans as a breakfast food, and they have the high blood levels of menaquinone to prove it.[42] The diets of other Asian cultures are lacking this source of K_2. This new data about what causes skin to sag show that K_2 plays a major role in maintaining a smooth, supple complexion.

The Way We Were

Photographic evidence from Dr. Weston Price's travels clearly illustrates that people following their ancestral, nutrient-dense diets aged much more gracefully than we do today. The photograph below is of a 90-year-old Polynesian woman, taken by Dr. Price circa 1930. Price described the women as having "splendid teeth and excellent facial and body build." Even after spending a lifetime under the South Pacific sun, this woman appears at least 20 years younger than someone her age and eating a non-traditional diet today would. Did K_2 play a role here? Modern research says yes, menaquinone fights skin aging and the emergence of wrinkles by protecting the elasticity of the skin in the exact same way it safeguards the elasticity of arteries and veins.[43]

© Price-Pottenger Nutrition Foundation, www.ppnf.org

Pseudoxanthoma elasticum (PXE) is an inherited condition that causes severe wrinkling of facial and body skin at an early age. Just as with age-related wrinkles, the elastic fibers of skin tissue of PXE patients become calcified, leading to skin laxity and not-so-fine lines. In 2007, researchers discovered the pathologic mechanism behind this disease: inactive MGP is abundant in the elastic fibers of PXE tissue samples. This uncovers yet another antiaging function for vitamin K₂: preventing the mineralization of skin tissue to keep your complexion firm and resilient. Moreover, the authors of the study were very specific: it is the high concentration of MGP that has not been activated that contributes to wrinkles.[44]

What will increase cellular production of MGP but leave it inactive? Taking large amounts of vitamin D (which increases MGP production) without also taking K₂. If you, like millions of others, have followed the good news about vitamin D by taking a supplement, you need to balance it with vitamins K₂ and A.

Vitamin K₂ for Healthy Veins

Nothing screams "old-legs!" like varicose veins. These uncomfortable, unsightly blood vessels affect about half of North American women, and up to 40 percent of men. The enlarged, knobbly veins can be flesh colored, dark purple or blue, and appear like twisted, bulging cords protruding from the skin surface. Varicose veins are most evident on the calf and thigh, where they are close to the skin surface, but it's important to understand that they can and do occur in other parts of the body—where they might not be as visible. Inflammation of the deep veins can lead to blood clots, which might break off and block the artery to the lungs. This is known as a pulmonary embolus, a potentially life-threatening condition. When an enlarged blood vessel occurs in the rectum or anus, it is called a hemorrhoid. Not life threatening, perhaps, but certainly very bothersome.

Considered esthetically unappealing, varicose veins also lead to pain that increases if you sit or stand for a long period. Heaviness in the legs, burning, throbbing, muscle cramps and swelling of the lower leg is also common. Sufferers may also experience itching around the vein(s). You should seek medical attention if skin ulcers appear near your ankle, as this is a sign of a severe form of vascular disease.

By definition, varicose veins are a problem with the cardiovascular system. Typically, after oxygen-filled blood circulates throughout the body, veins bring the oxygen-depleted blood back to the heart for replenishing. Inside healthy veins, one-way valves prevent gravity from drawing all the blood into our feet. When veins are damaged or not functioning properly, blood pools in the veins, leading to the visible bulges and lumps. Why does this happen? Very simply, as you age, your veins lose their elasticity and become lax, much like your skin loses its elasticity and begins to sag.

Traditional factors associated with varicose veins include aging, family history, hormonal changes (puberty, pregnancy, menopause, oral contraceptive use and hormone replacement therapy), obesity, leg injury and prolonged standing. A new risk factor is now being added to the list, but this is one you can really do something about. Yes, vitamin K₂ deficiency contributes to varicose veins.

Just as with atherosclerotic arteries and wrinkled skin, vitamin K₂–deprived MGP is prevalent in varicose veins. Smooth muscle cells from varicose veins show increased calcium deposition compared with healthy veins. Studies show that when MGP isn't activated—due to suboptimal K₂ levels—it contributes to the remodeling of the vein walls, causing bulging and distension.[45] Vitamin K₂ may be your best friend in winning the war on those achy, unattractive varicose veins.

More Help for Healthy Veins

In addition to ensuring adequate menaquinone intake, self-care for varicose veins includes weight loss (if warranted), regular exercise and elevating your legs while seated to offer symptom relief. Compression socks are also helpful, as they apply steady pressure to your legs, helping veins and leg muscles move blood more efficiently. Compression stockings are available at most pharmacies.

Other nutrients that improve the strength of blood vessel walls are flavonoids. These plant-based compounds are found in a wide range of fruits and vegetables, especially cherries, blackberries, onions and garlic, and in the white, outer membrane of citrus fruits. Try to include plenty of these foods in your diet, and look for a flavonoid supplement that includes hesperidin and diosmin. Horse chestnut extract is an excellent remedy for the symptoms of venous insufficiency, including pain and swelling of the legs. It won't address the underlying causes of blood vessel wall weakness, but it will temporarily support vein health.

Aging Well with Vitamin K₂

We're not meant to live forever, but maybe we are not doomed to degenerate as we age, either. A nutrient-dense diet complete with adequate fat-soluble vitamins is crucial in remaining healthy and vital after middle age, and slowing the aging process as much as possible. Taking vitamin K₂ may indeed be the most important step you can take for enhanced longevity, and its importance begins before we are born. Next we'll look at menaquinone-related conditions throughout life.

Even More Health Benefits of Vitamin K₂

V itamin K_2 helps prevent and treat most of the major and seemingly minor health concerns that plague the developed world. The benefits of menaquinone begin at and even before conception and extend throughout life. Exploring the essential role of K_2 in promoting optimal health at all stages of our existence uncovers a surprising link between outwardly benign health troubles and more grave concerns.

Vitamin K_2 for Diabetes

I mentioned earlier a vitamin K_2 connection that may prove to be the most exciting and unexpected benefit of this long-misunderstood nutrient: diabetes prevention. In 2007, groundbreaking research shocked the scientific community by revealing that our skeleton, via the vitamin K_2–dependent protein osteocalcin, has a significant impact on our body's production and sensitivity to insulin.[1] With that radical discovery, our perception of the skeleton makes a quantum leap from it being an inert scaffolding to it being a dynamic endocrine gland. Writing in the prestigious journal *Cell*, researchers explained that osteocalcin, produced within our bones, has the capacity to improve the body's glucose tolerance. And that makes vitamin K_2 critical for preventing an illness of epidemic proportions: insulin-resistant diabetes.

Insensitivity to insulin, otherwise known as insulin resistance, is the underlying cause of type 2 diabetes. Type 1 diabetes, also called juvenile-onset diabetes, is often diagnosed in childhood and occurs because the pancreas does not make insulin. Only about 10 percent of diabetics are type 1, and to date there is no known relationship between type 1 diabetes and vitamin K_2. Ninety percent of diabetics are afflicted with type 2, also called adult-onset or non-insulin-dependent diabetes. Within the decade, this obesity-associated lifestyle illness will become the biggest cause of disease and death in North America, and account for the biggest expenditure of health care dollars of all illnesses.

Insulin is a hormone secreted by the pancreas every time we eat, but much more so when we eat high-glycemic-index foods. High-glycemic-index foods are those that cause blood sugar levels to spike—sugar, white flour, bread, pasta and baked goods, for example. Insulin acts like a key that unlocks our body's cells to let sugar move from the bloodstream into the cells to provide us with energy and lower our blood sugar. However, when blood sugar and in turn insulin levels are too high too often because of a diet that's high in sugary, starchy or processed food, the cellular "lock" gets jammed up and the insulin "key" doesn't work so well. Over time, insulin becomes less and less efficient at opening the cellular doors to let sugar in: the overexposure to insulin causes cells to become resistant (insensitive) to it. Eventually, blood sugar levels remain high even though the body is producing plenty of insulin. The condition is referred to as "starvation in the midst of plenty" because the cells are starving for the energy from the sugar that circulates in the bloodstream but can't get in.

The second highest concentration of vitamin K₂ in the body is in the pancreas, the organ that produces insulin and governs blood sugar levels. In animals and humans, it seems that a lack of vitamin K negatively affects the pancreas's insulin production. For example, animal studies show that vitamin K deficiency negatively impacts glucose tolerance by slowing insulin response. Sugar stays in the blood longer and insulin levels are ultimately higher—the worst possible outcome for type 2 diabetes— when vitamin K is lacking. Basically, when researchers induce vitamin K deficiency, test animals develop type 2 diabetes.[2] Similar effects are seen in humans, although for ethical reasons human studies have focused on vitamin K intake rather than inducing vitamin K deficiency. Acute insulin response, the amount of insulin produced within 30 minutes of glucose intake, was found to be impaired in study subjects with low vitamin K intake.[3]

Only one week of vitamin K_2 supplementation in healthy, nondiabetic trial participants significantly reduces their two-hour postmeal insulin production by half.[4] Elevated insulin for hours after a meal is bad for diabetics, since it promotes insulin resistance. When insulin is functioning normally, it rises just enough in response to carbohydrate intake to get blood sugar levels back down again, then it declines so that it won't overexpose cells. The fact that K_2 supplementation lowers late (two-hour postmeal) insulin levels indicates that menaquinone helps insulin work more efficiently. This is certainly in keeping with the groundbreaking findings of the 2007 *Cell* study that revealed osteocalcin impacts insulin activity. Since osteocalcin improves insulin sensitivity and vitamin K_2 activates osteocalcin, it stands to reason that K_2 supplementation would improve insulin sensitivity.

Admittedly, changes in postmeal insulin levels are only an indirect measure of insulin resistance, and there are more specific markers to monitor this epidemic condition. Recent research from Japan shows that vitamin K_2 status is inversely related to insulin-resistant diabetes.[5] The more deficient diabetics are in vitamin K_2, the worse their results on several specific tests to determine blood sugar control and insulin sensitivity. These included the hemoglobin A_{1c} test, a measure of long-term blood sugar control, and the HOMA-IR test, a method that quantifies both insulin resistance and the function of pancreatic cells that produce insulin. Patients with more K_2-activated osteocalcin have better results on all of these tests, and better glucose tolerance. Other studies show that, although elevated levels of both carboxylated and undercarboxylated forms of osteocalcin are associated with improved glucose tolerance, only the K_2-activated form improves insulin sensitivity.[6]

Does taking vitamin K_2 improve insulin sensitivity? So far the intervention research examining the effects of taking a vitamin K supplement instead of just monitoring vitamin K levels have used only K_1, not K_2. This was a North American study where, unlike Japanese and European research that now focuses primarily on K_2, we're still fooling around with K_1. Be that as it may, the results are still interesting. First of all, 36 months of vitamin K_1 supplementation in people between 60 and 80 years of age slowed the progression of insulin resistance, although it didn't stop or reverse the disease.[7] I'm not knocking this benefit; a proven slowing of the advancement of such a serious disease is a good enough reason on its own to take vitamin K. But, given that carboxylated osteocalcin plays such an important role in insulin sensitivity, you'd expect a better result. Remember, K_1 isn't nearly as efficient at activating osteocalcin as K_2. As such, it's not surprising that vitamin K_1 supplementation had only a modest effect in this trial. Researchers should have looked at K_2 as well, but it does prime the pump for further studies using menaquinone.

The more curious finding from this particular study was that three years of vitamin K supplementation had a beneficial effect on insulin resistance in older men but not in older women. Why might that be? Recall from Chapter 4 that undercarboxylated osteocalcin increases in postmenopausal women, so they have greater requirements for vitamin K_2 at that time. It is highly likely that the treatment dose in this study was sufficient to benefit men but insufficient for older women with higher vitamin K_2 requirements. A similar trend is seen in cancer studies.

Again, it's not that women can't reap all the benefits of vitamin K₂ that men can. We just need more K₂ to meet our greater demands for bone health, in addition to cancer and diabetes prevention requirements.

Type 2 diabetes has a couple of infamous cronies, heart disease and osteoporosis. Do those conditions sound familiar? Insulin-resistant diabetics are highly likely to develop those vitamin K₂-deficiency diseases. Based on the emerging evidence from many scientific domains and the fact that type 2 diabetics are at the highest risk for two well-established menaquinone-deficiency diseases, osteoporosis and atherosclerosis, I recommend 240 micrograms of MK-7 to all of my diabetic, prediabetic and overweight patients.

Vitamin K₂ for Arthritis

Rheumatoid arthritis is a chronic, degenerative disease, affecting both children and adults, that is characterized by destructive inflammation of joints and surrounding tissues, including bone. Exactly what causes this form of arthritis is unknown, but it is thought it be autoimmune (when the body attacks itself) in nature. Might this be another manifestation of the Calcium Paradox? Adults with rheumatoid arthritis are at much higher risk of osteoporosis and cardiovascular disease, and the excess risk is not fully explained by traditional risk factors for those conditions.[8] That may be because vitamin K₂ deficiency has yet to be declared an official risk factor.

The classic bone and joint destruction seen in rheumatoid arthritis is caused by osteoclast activation. The synovium, the fluid-filled space within joints, contains the same cells that dissolve bone as part of the remodeling process. In rheumatoid arthritis, the cells cause more destruction than the body can repair, leading to joint decay. Clinical trials show that K_2, on its own or in combination with osteoporosis medication, keeps unruly osteoclasts in check to prevent joint damage in patients with rheumatoid arthritis.[9] In vitro studies—that is, lab experiments—show that K_2 inhibits the proliferation of other diseased cells seen in rheumatoid arthritis.[10] That makes menaquinone a promising new agent for the treatment of rheumatoid arthritis, with or without other anti-inflammatory drugs.

K_2 for Brain and Neurologic Health

Vitamin K_2 is a player in conditions, including Alzheimer's, that affect cognition and the central nervous system. The brain contains one of the highest concentrations of vitamin K_2 in the body after the pancreas, salivary glands and sternum, the cartilage that holds your ribs together and provides a convenient reservoir of vitamin K_2 for the heart and major arteries. Menaquinone's powerful ability to prevent free radical damage to neurons is one good reason for the brain to hoard this nutrient.

Oxidative stress causes brain cell death in numerous acute and chronic disorders of the brain, including stroke, transient ischemic attacks (TIA or mini-strokes) and any condition in which oxygen or blood supply to the brain is compromised, like sleep apnea. One neurological condition caused by severe oxidative damage to brain cells is cerebral palsy (CP). "CP" is an umbrella term encompassing many disorders of brain and nervous system function. In some people with CP, parts of the brain are injured by a lack of oxygen as a fetus, or during childbirth or infancy. If oxygen or blood supply to the brain is cut off, even for brief moments, at any point in early development, it can result in permanent brain damage from cell death. Fat-soluble-vitamin experts suggest that taking K vitamins during pregnancy could help prevent cerebral palsy in the newborn.[11] So we can also add this to the growing list of advantages of vitamin K₂ for pre- and perinatal health that we'll discuss shortly.

In addition to this remarkable antioxidant-like action, vitamin K₂ in the brain contributes to the production of myelin.[12] Myelin is the insulating material that forms a protective sheath around brain cells and nerves, much like the plastic outer coating on an electrical wire. It is essential for the proper functioning of the nervous system so neurological signals are conveyed correctly and efficiently. Multiple sclerosis (MS) is a condition in which patches of myelin on the brain and spinal cord become damaged. This may eventually lead to debilitating symptoms that affect the whole body, such as loss of coordination and muscle control, numbness, abnormal sensations, blurred vision, incontinence and more.

MS has long been linked to vitamin K₂'s fat-soluble friend, vitamin D. Studies show that the incidence of MS follows a geographic pattern. The disease is more common in areas of less sunshine, such as Canada and northern Europe. Also, fewer people with MS are born in November and more in May, suggesting that diminished sunlight exposure during pregnancy, which depresses mom's vitamin D levels, plays a role, too. Recently it was shown that some individuals with MS have a variation in their DNA that causes vitamin D deficiency early in life to lead to an autoimmune attack on myelin.[13] The implication is that ensuring children have an adequate vitamin D intake will help prevent MS in those with this genetic variant.

Will taking K₂ help prevent MS? Laboratory studies using animals is a preliminary step before entering into human clinical trials to determine if a particular treatment shows promise. Animals don't develop

MS, but research focuses on a similar illness in animals, called experimental autoimmune encephalomyelitis, or EAE. This disease is a widely accepted animal model of multiple sclerosis. Vitamin K_2 significantly reduces the severity of this MS stand-in when taken before the onset of symptoms.[14]

Research on vitamin K_2 and brain health is in its infancy. Since many of these early studies don't differentiate the form of vitamin K being studied, or look only at K_1, what's to say that K_2 really makes a difference here? In other words, can we achieve maximum brain-health protection just by eating green leafy vegetables? Probably not. K_2, not K_1, is the predominant form of vitamin K in the brain. Menaquinone outweighs phylloquinone in brain tissue by a ratio of 6 to 1.[15] In other words, the brain preferentially accumulates more than six times the amount of vitamin K_2 than it does K_1. Given the prevalence of vitamin K_2 in the brain, and the many mental and neurological conditions that are now implicated in vitamin K_2 deficiency, I wouldn't rely on the questionable conversion of K_1 to K_2 to meet our brain's need for menaquinone. The take-home message from research about the brain benefits of K_2 is that prevention is key: K_2 needs to be present before free radicals attack. Feed your brain with K_2-rich grass-fed foods, fermented dairy products, natto or supplements.

Vitamin K₂ for Cancer Prevention

Arguably the most promising recent research into the benefits of menaquinone pertains to cancer prevention. Since anticarcinogenic activities of vitamin K had been observed in both animal studies and human cell tissue (in vitro) studies in the laboratory, researchers were keen to uncover whether vitamin K intake has a real impact on cancer development in a practical sense. An initial level of evidence is epidemiologic studies. In other words, within a given population, do people who ingest more vitamin K_2 have a reduced chance of developing or dying from cancer than those who ingest less vitamin K_2?

The results of the first large study to address the association between dietary vitamin K intake and overall cancer risk were published in 2010. The European Prospective Investigation into Cancer and Nutrition (EPIC) found that the highest intakes of vitamin K_2 are associated with a reduced risk of developing cancer and overall death from cancer, with the latter reduced by about 30 percent.[16] This particular study followed more than 24,000 men and women between the ages of 35 and 64 for over 10 years. During that time, 1,755 cases of cancer were documented, 458 of which turned out to be fatal. Results showed that, independent of other cancer risk factors, people with the highest average intakes of vitamin K_2 were almost 30 percent less likely to develop cancer than people with the lowest average intakes. Dietary intake of menaquinone was more strongly inversely associated with fatal cancer than with cancer incidence. That reflects a pattern we'll see with many types of cancer. While vitamin K_2 deficiency doesn't cause cancer per se, it is connected to more aggressive and fatal malignancy.

Based on what had been seen in the laboratory, researchers hypothesized that the intake of both vitamin K_1 and K_2 would be associated with overall cancer incidence and mortality. What they actually found was that phylloquinone intake is not related to cancer risk. Only vitamin K_2, and not K_1, intake is related to cancer risk. This is remarkable because we know that a diet high in fruits and veggies helps protect against cancer—plenty of evidence supports that association. Since fruit and vegetables are rich in K_1, and we know that people who get their

5 to 10 servings a day have a lower rate of cancer than those who do not consume as many servings, we might expect to see a high intake of K_1 associated with lower cancer risk, but that is not the case. This isn't to say that you should oust produce from a cancer-prevention diet but, according to this research, it's not the K_1 in those foods that is protecting us from cancer. Only K_2 has that power.

Can you guess what was the main source of cancer-preventing vitamin K_2 for the EPIC study group? I'll give you a hint. Think back to the list of selected foods and their vitamin K_2 content in Chapter 3. Since this was a European study, our menaquinone chart topper, natto, didn't factor into the equation. The next highest food, goose liver pâté, is not a popular delicacy in the area of Europe where this large study was conducted. What's next on the list? Cheese. Researchers stated that the cancer-protecting dietary intake of menaquinone in this group was "highly determined by the consumption of cheese." Yes, cheese may be your secret weapon in the war against cancer.

Prostate Cancer

The authors of the EPIC study were quick to point out that the reduction of cancer risk with an increasing intake of menaquinone was more pronounced in men than in women. The main factor driving this finding is an inverse association between vitamin K_2 intake and the two biggest cancer killers of men, lung and prostate cancer. It is specifically the latter, prostate cancer, for which there is the fastest growing body of evidence of the cancer benefits of vitamin K_2. The first studies examining this relationship were published in 2008. Once again, menaquinone, but not phylloquinone, had an important impact on the progression of prostate cancer.

Although these studies found that men with the highest mena-quinone intake had overall lower risk of getting prostate cancer, the effect is not considered to be statistically significant. In other words, menaquinone doesn't seem to actually reduce the chance of getting prostate cancer. Although that sounds discouraging, it's not too surprising if you consider the nature of this particular form of cancer. Most men, if they live long enough, will develop cancerous cells within the prostate.[17] However, the majority of cases are slow growing and harmless, never extending beyond the relatively encapsulated prostate gland. Where K$_2$ intake does have a meaningful impact is on the risk of prostate cancer progressing to a life-threatening condition. More advanced-stage and high-grade prostate cancer is associated with menaquinone deficiency.[18] So, although developing innocuous prostate cancer may be more or less an inevitable aspect of aging for most men, menaquinone intake may make the difference between a potentially fatal cancer and one that goes unnoticed.

Study authors pointed out that menaquinone from dairy products had a stronger inverse association with advanced prostate cancer than did menaquinone from meat. Cheese was the greatest single food source of vitamin K$_2$ in the diet of most study participants, contributing 43 percent of total menaquinone intake. Meat was next, contributing 37 percent of total K$_2$ intake. When researchers looked more closely, only menaquinone from dairy products was associated with a significantly lower risk of advanced prostate cancer. This could mean one of two things (or both): meat contains other compounds that offset the cancer-fighting benefits of vitamin K$_2$, or the short-chain menaquinone (MK-4) in meat does not have the same ability to retard malignancy as do the long-chain forms of vitamin K$_2$ produced by bacterial fermentation, like MK-7. Further research will eventually sort this out. In the meantime, break out the Brie.

It's all very well and good to draw conclusions about menaquinone's effect on cancer based on diet questionnaires but, you might argue, maybe that's not the most accurate way to assess actual vitamin K_2 status. True, food-frequency questionnaires are a dietary assessment instrument with limited ability to determine absolute intakes. After all, absorption of dietary K_2 may vary from person to person. What about more specific markers of vitamin K_2 status? Do men with high levels of undercarboxylated osteocalcin (a specific marker of vitamin K_2 deficiency, explained in Chapter 6) have higher incidence of prostate cancer or prostate cancer mortality? Yes. Studies using a specific marker for K_2 deficiency show the exact same trend as in the population-based studies: menaquinone deficiency doesn't affect overall risk of developing prostate cancer, but it is significantly associated with advanced-stage prostate malignancy.[19]

Lung Cancer

After population-based studies, the next level of investigation examines the effect of menaquinone on cancer cells in vitro. K_2 continues to flex its cancer-fighting muscles here with another malignant biggie, lung cancer. Lung cancer remains the leading cause of cancer death in both men and women, accounting for almost one-third of all cancers. According to the American Cancer Society, an estimated 221,000 Americans will be diagnosed with lung cancer in 2011, and close to 157,000 will die of it. A similar trend is seen in Canada, in proportion to the population. Although smoking is still the major underlying factor for lung cancer, there are many types of malignancies that affect the lungs.

Researchers have tested vitamin K$_2$ against several major types of lung cancer cells and found that menaquinone suppresses cancer growth in all types of lung carcinoma.[20] Vitamin K$_2$ on its own is effective, but it also enhances the effects of the common chemotherapy drug cisplatin, providing a better cancer-killing effect than either treatment alone. Since menaquinone is a safe medicine, and without the prominent adverse effects of conventional anticancer agents, like bone marrow suppression, the evidence strongly suggests the therapeutic possibility of using vitamin K$_2$ to treat patients with lung carcinoma. Further laboratory studies show K$_2$ combats cancer in the liver, colon, stomach, breast, brain, nose, throat and mouth.[21]

Leukemia

Also in the K$_2$-conquers-cancer-in-vitro department are the many studies on menaquinone and leukemia. Leukemia (from the Greek *leukos*—white—and *haima*—blood) is cancer of the blood or bone marrow, characterized by an abnormal increase in white blood cells. "Leukemia" is an umbrella term encompassing a spectrum of diseases, with no single known cause. Although most leukemias are diagnosed in adults, it is the most common form of cancer in children and adolescents. U.S. estimates in 2010 predicted over 43,000 new cases of leukemia and approximately 21,800 deaths.

Scientists spotted the potent effect of vitamin K$_2$ on leukemia as early as the 1990s. Menaquinone induces apoptosis, or cellular death, for every type of leukemia cell tested.[22] It basically encourages the extra, abnormal white blood cells to self-destruct. The benefit is just as good for cancer cells cultured in the laboratory as for those isolated fresh from leukemia patient blood samples. K$_2$ also joins forces with vitamin

A for an even greater cancer-killing effect than either nutrient alone has. In contrast, as with other cancers, vitamin K_1 has no influence at all on leukemia.

While tricking any cancerous cells into committing suicide is impressive enough, K_2 doesn't wipe out all malignant cells. In other words, it doesn't cure cancer. Some types of leukemia cells seem to be particularly resistant to the effects of K_2, and menaquinone has only a mild to moderate ability to destroy these cells. However, in 2001, researchers discovered that menaquinone has another anticancer trick up its sleeve to deal with these rebels. Rather than obliterating the resistant cells, vitamin K_2 coaxes them to differentiate into harmless white blood cells. Cellular differentiation is a process by which less specialized cells develop into specific cell types. The level of differentiation is a measure of cancer progression; the more poorly differentiated the cells, the more developed the cancer. By spurring cancerous cells to either differentiate or die, K_2 provides a dual-action effect for patients with leukemia.

A number of striking case reports highlight the usefulness of K_2 as a cancer fighter, especially for leukemia. An 80-year-old woman with myelodysplastic syndrome or MDS, a kind of preleukemia, received an oral dose of 45 milligrams a day of K_2 as MK-4. After 14 months of treatment, her condition improved to the point that transfusions were no longer needed to maintain a normal white blood cell count. A 72-year-old woman diagnosed with acute leukemia achieved remission with a standard treatment, only to relapse eight months later. At that time, 20 milligrams a day of vitamin K_2 as MK-4 was added to the previous treatment. This resulted in the complete disappearance of cancerous cells after two months. Bone marrow analysis confirmed complete remission. A 65-year-old man whose preleukemia condition had progressed to leukemia was treated orally with 90 milligrams a day of MK-4. Within six weeks he experienced a significant decrease in abnormal blood cells and an increase in healthy cells. At 10 months the dosage was cut in half without side effects and the patient maintained good health without conventional chemotherapy.[23]

As with so many other health benefits of vitamin K$_2$, the leukemia-lashing effects are amplified when K$_2$ is combined with vitamin D$_3$. Given together, these vitamins intensify the differentiation activity, producing more normal, noncancerous cells. As an added bonus, the K$_2$-D$_3$ coalition helps prevent cytopenia, a dangerous lack of healthy white blood cells that often leads to life-threatening infection in patients undergoing cancer treatment.[24]

Liver Cancer

Once enough laboratory evidence and case reports accumulate for a given therapeutic benefit, the next step is intervention trials, studies that examine the effects of a treatment on real people. K$_2$ cancer research has reached this stage, albeit inadvertently. Early studies exploring the benefits of K$_2$ on bone density in women ended up uncovering another, unexpected benefit: vitamin K$_2$ prevents cancer in women with hepatitis.[25] Hepatocellular carcinoma, a cancer that originates in the liver, is a common complication of hepatitis B and C. This particularly deadly malignancy will kill almost all patients who have it within one year. Vitamin K$_2$ decreases the risk of developing liver cancer in women with hepatitis by delaying the onset of cancerous changes in liver cells. Vitamin K$_2$ also strongly induces the death of liver tumor cells, suggesting that menaquinone plays an important role in inhibiting tumor growth and invasiveness.[26] Clinical trials using K$_2$ for cirrhosis (liver disease) caused by hepatitis are so positive that I suggest that all hepatitis patients take a K_2 supplement.

❧   ❧   ❧

Astute readers might be wondering about the mechanism by which menaquinone exerts its anticancer effects. Scientists are wondering, too. Many kinds of malignant tumors, including prostate and breast cancer, produce the vitamin K_2–dependent matrix gla protein (MGP).[27] Since most of the early research on MGP and cancer does not distinguish between carboxylated and undercarboxylated MGP, it is premature to say exactly what MGP does in cancerous tissue. That K_2 deficiency is associated with more aggressive cancers strongly suggests that MGP left undercarboxylated is bad for cancer prognosis, just as it is bad for heart health. The controlling effect of vitamins A and D on MGP production is central to their cancer benefits. Certainly, the added benefit of K_2 and D_3 together hints at a relationship similar to that of bone and heart health, by which D_3 increases MGP so that K_2 can then activate it. Deciphering the riddle of vitamins A, D and K_2 may one day unlock a vast potential in cancer prevention and treatment. It is possible that by producing menaquinone-dependent MGP, malignant tissue itself contains the solution to controlling cancer.

Vitamin K₂ for Kidney Disease

The kidneys are our major organs of filtration, clearing wastes from the body and preventing toxins from building up in the blood. These paired organs also produce hormones that control other body systems, like blood pressure and red blood cell production, and they regulate the levels of minerals and fluid in the body. Many illnesses can ultimately hinder the kidneys' ability to cleanse the blood, with serious consequences.

Chronic kidney disease is a progressive loss of kidney function over time, which can be caused by a number of underlying conditions. Mild to moderate kidney function impairment often does not have symptoms and is diagnosed during routine screening of patients with other health conditions that predispose them to kidney problems, like diabetes. Chronic kidney disease is classified in five stages: stage 1 is the mildest, with few symptoms, and stage 5, previously called end-stage renal disease, is the most severe and fatal if left untreated.

Suboptimal levels of both vitamins K and D are prevalent in patients with chronic kidney disease.[28] The levels of vitamin K_2–deficient MGP increase progressively as chronic renal disease worsens, as does blood vessel calcification. Advanced kidney disease is invariably associated with vascular calcification and with renal osteodystrophy, a form of bone density loss that is distinct from osteoporosis.

Vitamin K_2 for Fertility

A major, recurring theme of Dr. Weston Price's work is the decline in fertility of parents and the decline in health of children that occurred when previously healthy people adopted a diet of modern foods. In the 1930s, health authorities were already very concerned about a diminishing reproductive capacity and a waning population owing to a smaller family size. Of course, many factors chip in to reduce family size. With higher levels of education and employment, marriage and babies are delayed, resulting in fewer babies. Today, however, up to 10 percent of couples are unable to conceive, even with reproductive assistance. The average male sperm count has been decreasing, especially in Western industrialized nations, by about 1 percent to 2 percent per year for decades, and now suboptimal male fertility with no apparent cause is relatively common.[29]

In both men and women, the sex organs are primary regulators of bone growth and bone density. As estrogen and testosterone spike at puberty, so does skeletal strength. Declining estrogen levels at menopause triggers bone loss for women, and a similar process happens to men because of declining testosterone levels starting in their late 60s. In males, at least, the relationship between bones and hormones is a two-way street. Osteocalcin, produced by the skeleton's bone-building osteoblast cells, induces the testes to produce testosterone. This stimulates sperm production and survival, having a fundamental effect on fertility. Osteocalcin-deficient male mice have significantly smaller testes and decreased sperm counts, and display 60 percent to 80 percent lower circulating testosterone levels compared with healthy litter mates. Matings between normal female and osteocalcin-deficient male mice produce smaller and less frequent litters than those between typical males and females. Osteocalcin binds to testicular Leydig cells, the body's key testosterone factories.

The dependence of osteocalcin production on vitamin D makes both vitamins D and K₂ essential to male fertility. Research on fertility and the benefits of vitamin K₂ is at its earliest stage of discovery. So far we have only an intriguing glimpse into another promising role for menaquinone, that of reviving diminished sperm production, but the implications are so important that this budding research deserves to be mentioned here. Chapter 7 reveals even more weighty evidence suggesting that vitamin E, a fat-soluble nutrient lost to grain feeding and flour refining, is central to reproductive efficiency.

Vitamin K₂ for Normal Facial Development

If you could promote straight, healthy adult teeth and avoid a future of painful, expensive orthodontic work for your unborn child simply by improving your preconception and prenatal nutrition, would you do it? Orthodontic appliances and wisdom teeth extraction are now so commonplace that nobody questions why our jaws are not wide enough to fit all the teeth that are genetically programmed to grow into them. Needing braces is considered just a normal phase of adolescent awkwardness for most kids. Dr. Weston Price made it clear that crowded, crooked teeth are not happenstance, nor are they due to heredity. Beautiful faces and healthy smiles are entirely the result of optimal prenatal nutrition, which provides plenty of fat-soluble vitamins, especially K_2.

The positive effect of optimal prenatal nutrition was arguably the most prominent theme of Price's work. The predictable result of inadequate nutrition before conception was clearly established as well. All around the world, the dentist photographed broad-faced indigenous parents with straight, healthy teeth and their offspring with identically wide, beautiful smiles—when, that is, the parents were following a traditional diet. Once the family adopted a diet of modern foods, the very same parents would produce children with a characteristic pattern of distorted, misaligned teeth (as seen in the photographs in Chapter 2). The severity of the deformity varied, depending on how long and how much modern food had become part of the diet. The more that life-preserving, smile-supporting nutrients had been usurped by nutrient-bereft processed foods, the more twisted teeth would result.

Price noted that almost every traditional culture he encountered had clearly defined rules about nutrition for both men and women who would become parents. Special foods, often fish eggs, were reserved for young adults who planned to marry and expand the family. In many tribes young hopefuls were not permitted to be married or even considered eligible for marriage until they had undergone a period of special nutrition. When those special foods for newlyweds and preconception practices were forsaken in favor of modern fare, the predictable pattern of facial changes appeared in the children.

Despite that chemical analysis showed that healthy traditional diets and special prenatal nutrition provided abundant fat-soluble vitamins, what's to say that a lack of K_2 specifically causes inadequate dental arches? After all, Price never narrowed down the dietary deficiency that causes crowded teeth to activator X or any single nutrient. His work in this area specifically emphasizes that a wide range of minerals and all the fat-soluble vitamins are necessary to produce healthy children. That's an important take-home point (and Chapter 7 explores the critical connection between the main fat-soluble vitamins). However, the latest research on vitamin K_2 and K_2-dependent proteins reveals that menaquinone is indeed the key to this dietary dental dilemma. Ensuring activation of all of the K_2-dependent proteins might be the most important step we can take to safeguard the smiles of our children and grandchildren.

Given the many starring roles osteocalcin has played so far, you might suspect it has some hand in this phenomenon, too. Surprisingly, normal facial development of the growing fetus, resulting in straight teeth later in life, has more to do with MGP—matrix gla protein. Preventing uncontrolled and inappropriate calcification is the key to healthy, attractive facial proportions. This is governed by, of all things, the prenatal nose.

Here's how it works. In the fetus, the developing nasal septal cartilage, the piece of cartilage that separates the two nasal cavities and forms your nostrils, is normally rich in the vitamin K$_2$–dependent MGP. Functional vitamin K$_2$–activated MGP is necessary to maintain growing cartilage in a normal, noncalcified state. When MGP is left inactive by a lack of vitamin K$_2$, premature calcification in the nasal cartilage stunts the growth of the face. Early hardening of the nose and jaw prevents these facial structures from reaching their widest, healthiest proportions. This results in an underdevelopment of the middle and lower third of the face, a condition technically called maxillonasal hypoplasia. Vitamin K deficiency in the first trimester of pregnancy results in maxillonasal hypoplasia in the newborn, with subsequent facial and orthodontic implications.[30]

It sounds extreme or monstrous, but it's a very common occurrence now. I guarantee you see it every day, to a greater or lesser degree, in the faces of friends, family, colleagues or maybe even your own reflection, if you have never had corrective orthodontic work like braces or cosmetic tooth extractions. All the children pictured in the photos of modern faces in Chapter 2 (page 32) display this to an extent. The degree to which this particular deformity has become pervasive in our culture is disguised by the ubiquity of orthodontic treatment. Along with the stunted development of the lower third of the face comes narrow dental arches that can't accommodate a full set of adult teeth. The orthodontic problems that lie ahead might not be obvious in children, since they have only 20 baby teeth. But, when the cuspids, the last of the adult teeth (not counting the wisdom teeth), erupt at age 12 or 13, they come in out of place. Very commonly, there just isn't room for these natural latecomers to develop on their own, so they are pulled out.

Cuspids are also called eyeteeth or canine teeth. They are the long, often pointy teeth on either side of the four front incisors. After the wisdom teeth, they are the most commonly impacted teeth. They are so often crowded that the American Association of Orthodontists recommends that a panoramic screening X-ray along with a dental examination be performed on all dental patients at around age 7 to determine if there will be problems with eruption of the adult teeth. If there isn't room for the eyeteeth to erupt normally, they will protrude behind or in front of the other teeth. Extensive orthodontic intervention is usually required to correct the problem.

Price clearly demonstrated that it was possible to avoid abnormal dental arches and promote healthy facial development of children with proper prenatal nutrition. Following the example of traditional tribes he visited, he prescribed special nutrition for his patients before and during pregnancy. This included "milk, green vegetables, sea foods, organs of animals and the reinforcement of the fat-soluble vitamins by very high vitamin butter [source of K₂] and cod liver oil."[31] This carefully chosen and highly nutritious fare provides protein, carbohydrates and healthy fat, with an abundance of minerals and water-soluble vitamins from the vegetables; vitamin D from the sea food, vitamins A and D from the milk, animal organs and cod liver oil; and plenty of vitamin K₂ from the oil of grass-fed butter.

The photos below illustrate one particular case that demonstrates the payoff of this special diet. The first child born to the family, shown on the left, has the classic narrow nostrils and crowded dental arch of gestational menaquinone deficiency. No special effort was made to reinforce the mother's nutrition during this pregnancy, and the girl's adult teeth came in displaying a typical variation of narrow facial form, that of slightly overlapping front teeth. Her pinched nostrils and small nose also caused the girl to breathe mostly through her mouth, a common manifestation of this syndrome.

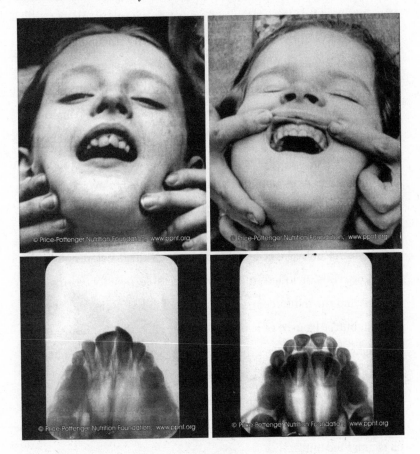

Before and during the second pregnancy, the mother's diet was enhanced with the food items mentioned above. Although the younger sibling, shown on the right, does not yet have her adult teeth, X-rays predict they will come in straight: this girl's dental arch is wide enough to accommodate all her adult teeth. It is unlikely that this is a coincidence. Aside from the fact that the nourishing diet during the second pregnancy provided abundant fat-soluble vitamins that were lacking during the first pregnancy, the second kid having straighter teeth than the first bucks a trend. It's not common for younger siblings to have better formed palates and dental arches than the firstborn—usually it's the other way around. The likelihood of having a dental arch deformity increases with birth order: second-, third- and fourth-borns have gradually escalating chances of having abnormal palate formation.[32]

This pattern of narrowing facial form with subsequent children was obvious among the traditional families that adopted modern foods. Firstborns, on the other hand, tended to have broad, well-proportioned faces, especially if the parents had not been eating modern foods for very long.

Is inadequate nutrition causing us to de-evolve?

In the photos in Chapter 2, you'll notice that the traditional faces with the widest, most attractive smiles (page 31) do not have perceptibly pointed canine teeth. Many of the grins showcase teeth that are so uniform, they seem to have been filed even. These perfect pearly whites are all natural, and one aspect of their appearance is almost conspicuous in its absence: fangs.

The word "atavism" means recurrence of an ancestral form. It is the reappearance of a lost character specific to a remote evolutionary ancestor. Examples of atavisms in animals include the appearance of reptilian teeth in a mutant chicken or vestigial hind legs in a whale. Human embryos in early development have tails and gills that normally disappear, but every now and then a child is born with the remnants of a tail stump. Atavisms occur because the genes for some previously existing features lie dormant in our DNA.

The appearance of pointy eyeteeth in the children of flat-toothed parents, triggered by nutritional deficiencies, is an example of atavism. Complete development of all teeth produces wide, even tooth arrangement. Incomplete tooth development brings forth the prominent canines seen in our primate ancestors.

Examples of the physical perfection documented by Price are rare today. There is one modern arena where you can still see many broad, perfectly proportioned faces, and that is professional sports. In particular, the NFL seems to have an abundance of wide, even smiles among its ranks. Not the quarterbacks so much, but the positions that require strength and speed, like linebackers, full backs and tight ends. Decathletes as well frequently sport square jaws and champion grins, and you'll notice a disproportionate amount of these excellent faces in any group of pro athletes. This statement is by no means backed by any kind of scientific validation; it's just a subjective impression. I'll bet money, however, that very few of these athletes got their straight teeth by way of orthodontic treatment. It is highly likely that the optimal nutrition that builds a healthy face facilitates physical perfection on many levels, producing strong, agile human beings.

The purpose of this seemingly trivial and unsubstantiated point is to assert that, ladies, you do not need to marry a professional football player to ensure healthy kids. You *could* do that, but without adequate prenatal nutrition for you and your hubby, those genes won't produce the robust offspring they should. Conversely, if you and your partner both wore braces as teenagers, you can potentially avoid that fate for your kids with the right nutrition. Our genes encode for physical perfection; we just need to provide the nutrition to let them reach their full potential.

Healthy facial development was so central to Price's research that it is represented in the logo of the Weston A. Price Foundation, the modern-day organization dedicated to disseminating the lessons of Price's work. The image depicts two wide ovals, representing the broad, healthy faces of people following traditional, nutrient-dense diets, on either side of a narrow oval. The narrow oval represents the typically narrow, underdeveloped face that is the product of an inadequate diet. The logo's wide, then narrow, then wide "faces" convey a hopeful message: we can restore the health of future generations by abiding by Mother Nature's wisdom and returning to highly nutritious diets.

Vitamin K₂ for Easier Labor

One of the outstanding changes Price documents in the traditional cultures at their point of contact with modern civilization is a "decrease in the ease and efficiency of the birth process."[33] What was once the most natural thing, childbirth, is now squarely in the domain of hospitals and medical intervention. Women who choose to have their children at home, even in the hands of a skilled midwife, are considered risk takers. This cultural phenomenon was borne of a medical necessity; somewhere along the way, childbirth became perilous business.

Sure, childbirth has always been risky business. Women and children have died during labor for as long as women have been giving birth. However, the vast majority of the human race, until very recently, came into this world without Pitocin, forceps or C-section. How did we possibly get by without those modern wonders? It would seem that, once upon a time, we didn't need them.

During his travels, Price was impressed with the "ease of reproduction" of the various traditional tribes he met. He gives the following example of an Inuit mother: "One Eskimo woman who had married twice . . . reported . . . that she had given birth to twenty-six children, that several of them had been born during the night and that she had not bothered to waken her husband, but had introduced him to the new baby in the morning."[34]

Okay, so maybe one extraordinary woman who is producing children like a T-shirt cannon doesn't prove anything. Is there any evidence to suggest that this reproductive prowess was a trend? Speaking of the members of a large Native reserve in Ontario, Price remarked, "The grandmothers of the present generation would take a shawl and either alone or accompanied by one member of their family retire to the bush and give birth to the baby and return with it to the cabin. A problem of little difficulty or concern, it seemed."[35] Things were different, however, for the contemporary generation of the same Six Nations tribe. Price goes on to say, "The young mothers of this last generation are brought to [the] hospital after they have been [in] labor for days. They are entirely different from their grandmothers or even mothers in their capacity and efficiency in the matter of reproduction."

Price gives multiple examples of the phenomenon, including this haunting report: "A similar impressive comment was made to me by Dr. Romig, the superintendent of the government hospital for Eskimos and Indians at Anchorage, Alaska. He stated that in his 36 years among the Eskimos, he had never been able to arrive in time to see a normal birth by a primitive Eskimo woman. But conditions have changed materially with the new generation of Eskimo girls, born after their parents began to use foods of modern civilization. Many of them are carried to his hospital after they had been in labor for several days."[36] And, with this, Price pinpoints the moment in history that we fell from grace, nutritionally speaking.

Apparently, the narrow face that personifies prenatal K₂ deficiency is also associated with a narrow pelvis. The fact that we are getting taller, on average, as a population, is probably not an indicator of improved nutrition but of worsening nutrition. A lack of fat-soluble vitamins produces longer, narrower bodies. Taller may be better to a certain point, but our genes encode for optimal body proportions, which are becoming distorted.

North American C-section rates are at an all-time high, with about 30 percent of children being born this way. There is modern evidence to uphold the proposition that a lack of fat-soluble vitamins interferes with the birthing process. Several studies show that women who are vitamin D deficient are four times more likely to deliver by cesarean section as women with higher vitamin D levels at the time of childbirth.[37] Where vitamin D is involved, we should question whether vitamin K₂ is also involved.

If pelvic narrowing was the only factor at play in progressively difficult labor, then we'd be in real trouble because there's not much you can do to widen a narrow pelvis. But Weston Price showed that it is possible to shorten labor times with the same nutritional treatment that promotes wider dental arches and straight teeth. Let's revisit the sisters shown in the photos on page 144.

This firstborn sister on the left came into the world at the end of 53 difficult hours of labor, after which the mother was debilitated for months afterward. The second daughter, whose birth was preceded by a special, nutrient-dense diet, popped out in only 3 hours and mom recuperated quickly afterward. Price remarks that labor difficulties are usually decreased and the strength and vitality of the child enhanced when mom has adequately reinforced nutrition, along the lines described above, during the formative period of the child.

Vitamin K$_2$ for Strong Bones

Osteoporosis prevention begins in childhood. Although the skeleton grows in both size and density until about age 30, up to 90 percent of maximum lifetime bone strength and density is acquired by age 18 in girls and by age 20 in boys. The amount of bony tissue present at the end of skeletal maturation is called peak bone mass. Experts agree that maximizing peak bone mass is an important way to prevent osteoporosis. The more bone tissue you can build by age 18 or 20, the lower your chances of succumbing to osteoporosis, and vitamin K$_2$ helps.

In both boys and girls, puberty is an especially dynamic period of bone development. Studies show that the rate of bone growth slows dramatically within three to four years after the onset of menses in girls and by age 18 in boys, who tend to hit puberty later. Because of that, a lot of bone health research is focused on how to optimize bone growth during this critical period. Teens have higher levels of undercarboxylated osteocalcin.[38] Not surprisingly, a better vitamin K status is associated with a more pronounced increase in bone mass in healthy kids 10 to 12 years of age.[39] Specifically, rather than looking at vitamin K intake from diet questionnaires, pediatric research has focused on a more direct indicator of vitamin K activity, the ratio of undercarboxylated (inactive) to carboxylated (vitamin K$_2$–activated) osteocalcin. In children nearing puberty, the more vitamin K$_2$ activates osteocalcin, the greater the increase in bone mineral content during that critical period.

Okay, so tweens with higher K_2 levels build better bones. Maybe they're just better nourished in general. Does supplementing with K_2 improve markers of bone quality in kids? Definitely. In healthy, prepubertal children, kids 6 to 10 years of age, modest supplementation with K_2 increases osteocalcin activation.[40] Clinical trials using only 45 micrograms of MK-7 for eight weeks show a decrease in undercarboxylated osteocalcin and an improvement in the ratio of active to inactive osteocalcin.

Vitamin K₂ for Dental Health

Of all the benefits of vitamin K_2, the one that shows the most promise and about which there is the least modern research is dental health. Working independently, Dr. Weston Price and his contemporaries showed that it was possible to not just prevent but to also heal active dental cavities with diet, and yet this research fell into obscurity. Ensuring adequate vitamin K_2 was a cornerstone of the nutritional protocol to treat cavities and it drills holes in our modern understanding of what really causes cavities and how to treat them.

The tooth is made of four parts (see the diagram below). The soft innermost layer is called the pulp. It houses blood vessels connected to the body's circulatory system and sensitive nerves. Below the gum line is the tooth's root; above the gum line is the crown. The pulp is surrounded by dentin, a calcified, bonelike matrix made up of millions of tiny, closely packed tubules. In the root, the dentin is covered by cementum, a thin layer of mineralized tissue. In the crown, the dentin is covered by enamel, the white portion of tooth we can see.

Anatomy of a Tooth

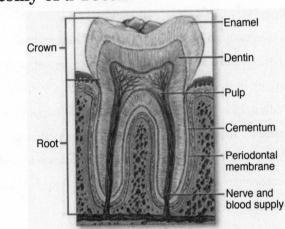

Of the three calcified tissues, enamel, dentin and cementum, dentin is unique for a couple of reasons. Unlike enamel, which is formed largely in the womb, dentin continues to form throughout life. Under the right conditions, dentin production is stimulated in response to triggers like tooth decay and even chewing. Odontoblasts, cells very similar to the bone-building osteoblasts, line the surface of the pulp just beneath the dentin and continually produce new dentin. Dentin is also unique because it produces the vitamin K_2–dependent proteins osteocalcin and MGP (matrix gla protein).

Tooth decay starts from outside the tooth. Cavity-causing bacteria produce acid that slowly eats through the enamel, then quickly eats through the more porous dentin. Traveling along the tiny dentin channels, bacteria quickly reach the pulp, which may become infected even before the tooth decay penetrates all the way through the dentin. Regular dental checkups and X-rays that spot cavities early limit the progression of a cavity. Drilling out the decay and replacing the lost tooth matter with a filling effectively seals out bacteria and stops the cavity from growing. But that doesn't prevent the process from starting all over again in another tooth, or even in another part of the same tooth.

There are a few different microorganisms involved in tooth decay, namely the *Streptococcus* species and certain strains of *Lactobacillus acidophilus*. If that last name sounds familiar, it's because these bugs are considered to be probiotics—friendly, helpful bacteria—in other parts of the body. In the intestines, this species helps digest food and boost immunity. They are found in yogurt and probiotic supplements. Yes, we've been paying money for the bacteria that cause our teeth to rot. If those bacteria are so helpful elsewhere, why are they harmful in our mouth? What induces bacteria to attack teeth?

According to the conventionally accepted understanding of tooth decay, cavities happen when foods containing sugars and starches, such as bread, soda pop, cookies, candy or even milk, are frequently left on the teeth. Mouth-dwelling bacteria thrive on these foods, producing acid that, over time, destroys tooth enamel, resulting in decay. Thus, eating high-carb foods and not brushing your teeth causes cavities because of this localized reaction. Good oral hygiene will reduce bacteria, while dietary changes reduce what they feed on. This is the chemicoparasitic theory of tooth decay.

This narrow view of what causes cavities has us playing a losing game of catch-up with tooth decay—and it doesn't leave room for the most effective and fundamental approaches to preventing cavities way before they start. Even with a low-sugar diet and regular brushing, flossing and professional cleaning, cavities happen. It is impossible to keep the mouth free of bacteria with dental hygiene. More to the point, Price found that "many primitive races have their teeth smeared with starchy food almost constantly and make no effort whatsoever to clean their teeth. In spite of this they have no decay."[41] Something else protected these people from both the bacteria and their cavity-causing activity: vitamin K$_2$.

Price observed that people with active tooth decay had high levels of *Lactobacillus acidophilus* in their saliva, averaging around 323,000 microorganisms per milliliter. After treating his patients with vitamin K$_2$–rich butter oil, Price's special concentrate of butter from grass-fed cows, the average bacteria content dropped to 15,000 bugs per milliliter of saliva, a reduction of 95 percent.[42] In some cases, the bacteria disappeared completely. The almost complete elimination of bacteria was typical in "many hundred[s] of clinical cases in which dental caries [are] reduced apparently to zero, as indicated by both x-ray and instrumental examination."

The addition of dietary K$_2$ changes the quality of saliva in another surprising way that fights tooth decay. The saliva of patients who have cavities tends to rob the teeth of minerals, according to another elegant experiment performed by the maverick dentist. When saliva from patients with active tooth decay was mixed with powdered bone or tooth chips, minerals moved from the tooth or bone into the saliva. The experiment was repeated with saliva from the same patients after they were treated with vitamin K$_2$. Then, minerals moved from the saliva into the bony tissue.

After the pancreas, vitamin K$_2$ in humans exists in the highest concentration in the salivary glands. When rats are fed only K$_1$, nearly all of the vitamin K in their salivary glands exists as K$_2$.[43] Vitamin K$_2$ accomplishes two things here. It reduces the number of cavity-causing bacteria, and it provides dentin with the menaquinone needed to activate MGP and osteocalcin. Once those proteins are activated by K$_2$, they develop "claws" that grab onto calcium to deposit it where it's needed. That mechanism alone could explain the tendency for minerals to be drawn into tooth tissue in the presence of vitamin K_2–rich saliva.

Once Price recognized the value of vitamins A, D and K_2 in treating tooth decay, he largely stopped drilling and filling teeth, except in cases where pain from a large, open cavity called for a temporary filling. Instead, he used a combination of high-vitamin cod liver oil (source of vitamins A and D) and grass-fed butter oil (source of K_2) as the foundation of his protocol for healing cavities. This protocol not only stopped the progression of tooth decay but completely reversed it by causing dentin to grow and remineralize, sealing what were once active cavities.

Let me be clear that I am not advocating that you give up dental care and self-treat your family's tooth decay with Price's method, as described in these pages. A nutritional protocol for treating tooth decay should be overseen by a patient and informed dentist. I'm also not suggesting we all give up brushing and flossing. Even Price admitted that "of course everyone should clean his teeth, even the primitives, in the interest of and out of consideration of others."[44]

The Bigger Picture

There's one last stumbling block for the conventional concept of what causes cavities, one that has far greater implications than dental health alone. If tooth decay were just a matter of chemicals and parasites, then, as adults, we would continue to develop new cavities the way we did in childhood. But it doesn't generally happen that way. If you had cavities growing up, like me, you were probably in your 20s when you had your first "good" checkup, and there were few new cavities thereafter. That's because the incidence of new cavities that penetrate the enamel peaks around puberty and slows down in adulthood.[45]

Something else, besides raging hormones, peaks around puberty, as you might recall. The rate at which bone tissue accumulates maxes out at this time, and teens have the vitamin K_2 deficiency to prove it. Kids and teens get more cavities because their growing bodies have a greater need for bone-building nutrients. If a diet of modern, mass-produced food isn't supplying enough nutrients during this critical growth period, there won't be any to spare for the teeth, resulting in decay. Remember the triage theory of aging, discussed in Chapter 4? You can bet that if the teeth are suffering from nutrient deficiencies, peak bone mass isn't going to be optimal, either. A bad checkup shouldn't just mean another filling and a lecture about dental hygiene. When our children develop tooth decay, it is a wake-up call that their diet isn't providing enough minerals to fill their teeth and growing bones, or enough fat-soluble vitamins, especially K_2, to keep those minerals in place.

Price wasn't the only doctor who figured out that dental cavities could be healed with diet. Other researchers of his day, notably Sir Edward Mellanby, physician and codiscoverer of vitamin D, together with his wife Lady May Mellanby, published protocols of specific diets high in fat-soluble vitamins that kept cavities in check. The program was so effective that they considered the problem of "retarding or arresting dental caries by dietetic measures" to be solved.[46] Their only concern was why the program didn't work for *every* child—why some children responded better than others. They determined that this was due in part to the consumption of phytic acid, found in whole grains, an issue we'll examine in Chapter 8. More to the point, these scientists observed that there was a difference in the structure of the teeth, determined before birth, in the children who didn't respond completely to the cavity-curing diet. It turns out that the real work of preventing tooth decay begins before we are born.

Our teeth are formed in the womb. Primary (baby) teeth start to develop between the sixth and eighth week of pregnancy, while the permanent adult teeth that you won't see for another 6 to 12 years form during the second trimester. Not surprisingly, the same nutrients that govern the development of the face and dental arches govern the development of the teeth that form within those arches. Developing teeth contain both MGP and osteocalcin, meaning they need vitamin K$_2$ to form properly.

For the researchers of the early 20th century, completely preventing and treating tooth decay in every child using dietary methods was an achievable goal, and beginning prevention before birth was just logical. These scientists concluded, "Since the majority of human teeth in this country are defective in structure it would seem that one method of reducing the incidence of dental caries should be the production of teeth of good structure by suitable diets for mothers during pregnancy and lactation and for their offspring during the period of tooth calcification. This is essentially the prophylactic method of attacking dental caries."[47]

A New Understanding of Tooth Decay and Oral Health

Our teeth are designed to heal. Eroded, decalcified dentin renews itself when the right nutrients are plentiful. Tooth decay itself stimulates the production of new dentin, as does the mechanical stimulation of chewing. That's how people in some traditional cultures were able to eat foods that, over a lifetime, wore their teeth down to the gum line without causing any tooth decay. Similarly, some tribal groups filed their teeth to dramatic points for esthetic reasons without suffering any ill effects. Our modern diets don't contain any food abrasive enough to wear down teeth, but when decalcification from decay exposes the inner pulp of a 21st-century tooth, it invariably becomes infected by bacteria. Modern teeth lack the vitamin that generated a protective shield in people eating traditional, nutrient-dense diets: vitamin K$_2$.

In many ways, the loss of K$_2$ has impacted women more than men. In particular, women get more cavities than men.[48] It wasn't always this way. Females experienced a more rapid decline in dental health than did their male counterparts as humans made the transition from hunting and gathering to agricultural lifestyles. Is that because we're not brushing and flossing as well? No, if anything women are more conscientious than men when it comes to oral hygiene and dental care.[49] Women get more cavities for the same reason that we fall behind in the cancer and diabetes benefits of vitamin K$_2$: we are more susceptible to vitamin K$_2$ deficiency. The demands of reproduction and the nature of female sex hormones all chip away at our K$_2$ status—until we recognize the warning signs for what they are and take action.

This "new" understanding of tooth decay as nutritional deficiency frees women and men to take effective action in preventing and treating cavities, with far-reaching implications. Weston Price was primarily interested in activator X because of its powerful ability to control dental cavities. He also recognized that the significance of oral health extends far beyond the mouth. When bacteria from the mouth make their way through the dentin tubules to reach the tooth's pulp, they gain access to the bloodstream. This infection starts the degeneration of organs and tissues in other parts of the body, like the heart.

One principle of traditional Chinese medicine is, "The heart opens onto the mouth." According to this ancient wisdom, heart health is reflected in the health of the mouth. The Chinese were on to something. Modern evidence shows that people with gum disease are twice as likely to suffer from coronary artery disease.[50] Gum (periodontal) disease is characterized by an accumulation of soft plaque, which calcifies to tartar, around the teeth. This mirrors the buildup of arterial plaque of heart disease. Of course when calcium deposits on our teeth we just get it scraped off, but it is a forerunner of many other vitamin K₂ deficient conditions.

For example, periodontal disease is an early complication of diabetes. The connection between these two concerns is so tight scientists recently declared that dentists can effectively identify people with undiagnosed diabetes and prediabetes just with routine dental exams.[51] Medical and public health experts emphasize the importance of early detection of diabetes, since this can limit the development of severe complications and about one quarter of people with diabetes don't know they have it. Doctors would do their patients a favor by recommending vitamin K₂ to benefit both conditions, instead of just treating each concern separately.

Additional studies point to a relationship between gum disease and stroke. Stroke is a manifestation of cardiovascular disease, one that affects the brain. Stroke is sometimes called a "brain attack," a name that conveys its similarity to heart attack. Even with conventional risk factors for cardiovascular disease taken into account, dental infections are associated with both stroke and heart attack.[52] The most surprising finding is that gum disease is a stronger indicator of total mortality risk (death from any cause) than coronary artery disease.[53] Other research shows a strong association between bone loss from gum disease and fatal coronary heart disease and stroke.[54] Does this pattern sound familiar?

A major shortcoming of most of the science on the periodontal disease–cardiovascular disease link is that it focuses on a single risk factor for gum disease, oral hygiene. This suggests that the only way to beat gum disease is by brushing and flossing. We're back to the chemicoparasitic model here. Of course, oral hygiene is important, but, as with cavity prevention, it does not address underlying susceptibility. Correcting fat-soluble-vitamin deficiencies does. New evidence shows vitamin D deficiency causes periodontal disease.[55] Although we have few clues about the mechanism, and likely there is more than one, Price's experiments with K_2 and dental bacteria propose yet another realm in which fat-soluble vitamins unite for our benefit.

The amazing interplay between vitamins K_2, A and D flushes out the story of how menaquinone optimizes our well-being. That account is coming soon, but not before we look at how to discern if you need to improve your K_2 status. I probably don't have to talk you into eating more cheese. However, since we've established that you can be silently deficient in K_2, how can you tell if you should be adding this nutrient to your pile of supplements, in addition to eating more K_2-rich food? Chapter 6 explains the ins and outs of K_2 testing.

Measuring Your Vitamin K₂ Levels

M ost of the evidence we've covered so far about the risks of vitamin K_2 deficiency, and the benefits of a menaquinone-rich diet, is based on using some kind of laboratory assessment of K_2 intake adequacy. Since the major repercussions of K_2 insufficiency tend to be asymptomatic until, say, a catastrophic heart attack or hip fracture, it might be helpful to have an earlier warning system for a deficiency of the most important nutritional factor contributing to osteoporosis and atherosclerosis. On the other hand, vitamin K_2 deficiency is common and there aren't any known ill effects of taking K_2, so maybe we should all just do that. Are you getting enough vitamin K_2? Do you need a test to determine that? Here we look at assessing whether your diet is providing adequate menaquinone to meet your health needs and whether you should bother trying to find out your vitamin K_2 status.

Spoiler alert. I might as well tell you up front that as this book goes to print, the most useful tests to measure vitamin K_2 status are not yet readily available for diagnostic purposes in North America. If that's the case, why bother with this chapter? Because it won't be long until really worthwhile K_2 tests are obtainable, and this information will help you know what to look for. Also, other diagnostic tools that measure the results of K_2 therapy are out there, so we'll cover those, too.

The demand for blood tests evaluating the levels of another fat-soluble nutrient, vitamin D, has skyrocketed in the last five years. With greater appreciation of the many health benefits of vitamin D, and awareness that deficiency is rampant in northern latitudes, patients and physicians are seeking vitamin D testing in droves. The practice is simple enough. Draw some blood and send it to the lab to measure how much 25-hydroxy vitamin D is in the sample. If your level is around 100 nmol/L (nanomoles per liter), you're doing well.[1]

Testing doesn't work the same with K_2. The little K_2 that we do store in our body tends to be stowed away in organs and tissues like the pancreas, salivary glands, sternum and brain. As such, detectable levels of menaquinone don't circulate freely in our bloodstream, as with vitamin D. Vitamin K_2 levels can't be measured directly, so we have to take a more functional approach by measuring how much vitamin K_2–dependent protein is activated by K_2. Scientific trials use either osteocalcin or MGP (matrix gla protein) carboxylation as a marker for K_2 status.

Undercarboxylated Osteocalcin (ucOC) Tests

Osteocalcin is secreted by osteoblasts and odontoblasts. These bone- and tooth-building cells use the protein to draw minerals, especially calcium, into the collagen matrix of bony tissue. Osteocalcin is the major noncollagen protein found in bone, and it was the first vitamin K–dependent protein to be discovered that did not play a role in blood clotting.

In addition to building bone density, osteocalcin acts as a hormone. It causes insulin-producing beta cells in the pancreas to release more insulin, and at the same time directs fat cells to release the hormone adiponectin, which increases sensitivity to insulin. Osteocalcin also affects male fertility by enhancing the synthesis of testosterone, which boosts sperm production.

Osteocalcin sometimes goes by the less popular handle of "bone gamma-carboxyglutamate protein" (BGLAP), or just "bone gla protein" (BGP) for short. It's that "gla" (gamma-carboxyglutamate) portion that makes this protein vitamin K_2 dependent. K_2's effect on that part of the osteocalcin protein forms the basis of the most common diagnostic test for K_2 sufficiency. In its intact form, when first synthesized by osteo-blasts, osteocalcin contains three so-called gla residues. These amino acid components protrude from the protein and can be modified to bind calcium. When vitamin K_2 is present, it carboxylates (activates) osteocalcin, and those gla residues undergo a detectable change. When vitamin K_2 is lacking, some or all of those gla residues remain unchanged and osteocalcin remains unable to bind calcium. The undercarboxylated osteocalcin (ucOC) is distinguishable from the activated form, so this makes it a convenient marker for vitamin K_2 status. When K_2 is low, ucOC is high, and vice versa.

We're in the early days of K_2 investigation. Different types of ucOC tests are used in research, which is why it's sometimes hard to compare the results of one study with another. Of the two main methods for judging K_2 sufficiency, one is used primarily in research settings, and the other is more readily available to the public via their doctors.

The ucOC Ratio Test

The first, more academic evaluation of K_2 status compares the amount of undercarboxylated osteocalcin to the total amount of osteocalcin in a blood sample; the results are expressed as a ratio or percent. This is a logical way to evaluate if K_2 intake is adequate to meet osteocalcin demand. An analysis of the various available methods for determining vitamin K status concluded that measuring ucOC and expressing that finding as a ratio or percent of total osteocalcin is the most accurate way to proceed.[2]

Unfortunately, ucOC ratio tests are not yet commonly available to commercial laboratories that service health care providers, thus making them unobtainable to the public. It's too early to discuss what an ideal result would be on this test, since it depends on whether under-carboxylated osteocalcin is being compared to total osteocalcin or only to carboxylated osteocalcin. Those would both be valid measurements, and a healthy ratio depends in part on the test being used.

The Absolute ucOC Test

After ratio testing, the ucOC test is your next best bet. This test measures the absolute amount of osteocalcin circulating in the blood that is not activated by vitamin K_2. It does not compare ucOC to total or carboxylated osteocalcin, making it only a rough estimate of true vitamin K_2 sufficiency. The drawback to this test is that factors unrelated to vitamin K_2 status affect osteocalcin production, so they may affect the results. More or less undercarboxylated osteocalcin could be hanging around just because there's more or less total osteocalcin available to be carboxylated. The researchers who developed the test provide a reference range (the average upper limit for healthy people), but the test would need to be applied to a large population to confirm the validity of the range.

Elevated levels of ucOC, indicating a vitamin K_2 deficiency, are associated with lower bone mineral density and higher risk of hip fracture. Levels of ucOC increase in teenagers, reflecting a greater need of vitamin K_2 for their growing bodies, and levels spike again at menopause when the bone-boosting effects of estrogen are withdrawn. This effect is seen in women who undergo natural menopause and in those whose ovaries have been removed surgically.

What is a passing mark on an ucOC test? Remember, the more undercarboxylated osteocalcin you have, the more deficient you are in K_2, so the lower your ucOC score, the better. The average result for healthy people, according to the researchers who developed the test, is around 3.6 ng/ml (nanograms per milliliter).[3] Levels of ucOC tend to spike in teenagers and menopausal women, so this rough test does reflect the same trends of vitamin K_2 deficiency we see using ratio measurements. Some studies noted diminished bone density with ucOC as low as 1.65 ng/ml. Keeping in mind that apparently healthy people are at risk for long-term complications of having undercarboxylated osteocalcin, we should probably be striving for a value below 1.6 ng/ml.

Corticosteroids and ucOC testing

Corticosteroid treatment may interfere with the results of any kind of osteocalcin testing, including the ratio tests. Anti-inflammatory steroid medications, like prednisone, are notorious for their bone-density-lowering side effects. Corticosteroids lower the level of both osteocalcin and ucOC, with or without vitamin K_2 supplementation.[4] That means that steroids confound the usual association between vitamin K activity and ucOC levels, so osteocalcin testing may not be a reliable indicator of vitamin K status if you are taking corticosteroids.

The only company currently offering the ucOC test is Metametrix. Since the test is not the desirable ratio assessment, its value is limited. Only a licensed health care professional may order the test, which is called the Vitamin K Assay in the company's catalog. If your doctor doesn't currently offer it, the company's user-friendly website has a downloadable form that you can bring to your doctor so he or she may order it. Alternatively, Metametrix provides a clinician referral service that will connect you to a practitioner in your area who works with this lab.

The Serum Osteocalcin Test

Another test, the name of which is confusingly similar to the one that gauges K_2 activity, is much more commonly available than the ucOC, so watch out you don't accidentally get this if what you really want is to measure your menaquinone status. It's the serum osteocalcin test, also called bone gla protein, or BGP, test. The serum osteocalcin test measures the amount of osteocalcin in circulation. Osteocalcin levels are a biochemical marker, or biomarker, for the bone formation process. Serum osteocalcin levels are correlated with bone diseases like osteoporosis, so your doctor might suggest this test to evaluate your risk of osteoporosis or to assess the effectiveness of osteoporosis treatment. Clinical trials use osteocalcin levels as a preliminary biomarker of the effectiveness of bisphosphonates, a class of drugs used to treat osteoporosis.

The osteocalcin test is *not* the same as the undercarboxylated osteocalcin test, and it won't measure your vitamin K_2 status. Osteocalcin testing measures only how much of this vitamin K_2–dependent protein is present in the blood, not how much of it has been activated by K_2. Measuring osteocalcin levels isn't a good substitute for measuring ucOC levels because many factors unrelated to vitamin K_2 can affect the test. Osteocalcin levels may increase or decrease, but that doesn't tell you how much of the osteocalcin that is there is activated by K_2. Osteocalcin levels don't correlate with disease risk for osteoporosis, heart disease or cancer the same way ucOC levels do. Osteocalcin assessment is a useful tool in managing certain health conditions and may be indicated for you, it just can't be used to judge whether you are getting enough menaquinone.

Should You Be Tested?

Given that the only truly meaningful undercarboxylated osteocalcin testing is currently unavailable, vitamin K_2 deficiency is common and there are no known side effects of taking vitamin K_2, the average person needn't bother with undercarboxylated osteocalcin testing. Just focus on a diet that includes grass-fed foods, fermented dairy products and natto, or take a K_2 supplement if your daily diet is lacking menaquinone-rich foods. For you keeners out there who just love empirical evidence, or for clinicians who suspect vitamin K_2 deficiency is affecting their patient's health, the undercarboxylated osteocalcin test is an option.

Bone Density Scanning

While we're waiting for better ucOC tests to arrive on the scene, monitoring bone mineral density (BMD) is something to consider, as it gauges one of the desired benefits of taking vitamin K_2. Vitamin D intake, calcium intake, hormones and other factors all influence bone density, so bone density testing is not a measurement of K_2 status only. However, K_2's effects on bone health are noticeable with BMD testing, which measures a crucial, though by no means unique, predictor of fracture risk.

Dual-emission X-ray absorptiometry (DXA or DEXA) is the most common test for measuring bone mineral density. DXA is an accurate tool for diagnosing osteopenia or osteoporosis, and periodic scans can be used to monitor osteoporosis treatment, including vitamin K_2 therapy. A DXA scanner is a machine that produces two X-ray beams, each with different energy levels; one beam is high energy, the other is low energy. The amount of X-rays that pass through the bone is measured for each beam—it will vary depending on the thickness of the bone. The bone density is then calculated based on the difference between the two beams. DXA doesn't measure the density of every bone in the body but usually only the hip and spine, since factures in those areas are the most dangerous.

DXA scans are used to measure bone mineral density because they are more accurate than regular X-rays. A person would need to lose 20 percent to 30 percent of their bone density before it would show up on an X-ray. DXA scans also involve less radiation than other forms of bone mineral testing. Although two X-ray beams may sound like a heavy dose of radiation, a DXA scan delivers less radiation than a chest X-ray.

Based on the result of the scan, your BMD will be expressed as a T-score. Your T-score is a comparison of your BMD to that of a healthy 30-year-old of the same sex and ethnicity. A normal T-score is −1.0 or higher, osteopenia is defined as between −1.0 and −2.5 and osteoporosis is defined as −2.5 or lower, meaning BMD is two and a half standard deviations (or more) below the average for a 30-year-old woman or man.

Bone mineral density isn't everything

Learning you have osteoporosis can be an emotional shock, especially if you feel confident you've been living a healthy lifestyle. Your DXA results are not the ultimate predictor of whether you will eventually break a hip. The diagnosis of osteoporosis in postmenopausal women is based only on bone density testing by DXA scanning, but factors other than bone density contribute to your probability of having a fracture or break. BMD is an important factor but not the only determinant of fracture risk. Most patients and some physicians are surprised to learn this.

Other aspects of skeletal micro-architecture that aren't measured by DXA scans contribute to bone strength. Your age, personal history of falls and previous fractures also significantly influence the odds of suffering an unlucky break in the future. The most current clinical guidelines recommend physicians move away from prescribing osteoporosis medications based solely on DXA results and include other considerations to better target osteoporosis therapy.[5] Fortunately, since increasing your K_2 intake has no known downside, you don't have to wonder whether or not you should do it.

The recommended frequency of follow-up DXA scans is controversial. Some doctors recommend the scan be done every year or two. However, bone mineral density typically changes less than 1 percent per year, which is less than the margin of error of a DXA machine. Changes of less than 4 percent in the vertebrae and 6 percent at the hip from test to test can be due to the precision error of the scan. For that reason, too frequent repeat scanning may not reflect meaningful changes in bone density. Also, although the more porous the bone, the more prone it is to breaking, there isn't a perfect correlation between increases in bone density as measured by DXA with decreases in fracture risks with treatment. So don't be too emotionally invested in small changes in your T-score.

Strontium supplements are gaining in popularity for bone health, and they will affect the results of your DXA scan. Strontium is mineral that is absorbed into bones in a similar fashion as calcium to increase bone density. DXA scans are usually read based on calcium as the main component of your bones. If you have been taking strontium, it will be present in your bones and affect the way X-rays pass through your bone tissue. Strontium supplements may artificially elevate your BMD results by up to 50 percent, so let the DXA technician know if you have been taking strontium, at what dose and for how long.

What effect will K_2 have on your DXA scan? Although studies show that K_2 deficiency is associated with lower BMD, clinical trials to determine the specific bone density payoff of K_2 supplementation are ongoing. If your K_2 intake is sufficient to meet your bone health needs, you can expect a maintenance of your lumbar spine BMD. K_2 also reduces hip fracture in ways that aren't measurable with BMD testing.[6]

Should You Be Tested?

If you are a woman over the age of 65, you should get a DXA scan. In addition, postmenopausal women under 65 years who have risk factors for osteoporosis other than menopause (such as a previous history of fractures, low body weight, cigarette smoking or a family history of fractures) should be tested. Finally, women and men over 50 with strong risk factors (listed below) should discuss the benefits of DXA scanning with their doctor.

Risk Factors for Osteoporosis

- personal history of fracture as an adult
- low body weight or thin body stature
- cigarette smoking
- use of corticosteroid therapy for more than three months
- impaired vision (increase risk of falling)
- estrogen deficiency at early age
- dementia
- poor health/frailty
- frequent falls
- low calcium intake
- low physical activity
- alcohol intake of more than two drinks per day
- thyroid disease
- rheumatoid arthritis
- excessive caffeine consumption
- use of oral contraceptive

Tracking Your Plaque

There is another indirect assessment of vitamin K_2 activity that gets straight to the heart of the Calcium Paradox. Coronary artery calcium scoring is a specialized type of ultra-fast X-ray imaging that measures the presence and amount of calcium buildup in the arteries that supply blood and oxygen to your heart. If your main goal of K_2 supplementation is reducing your risk of heart attack, this is an important test to take, to measure that risk.

Coronary artery calcium (CAC) scoring, also called calcium score or heart scan, is a technique that uses computed tomography (CT) scanning technology to quantify the volume and density of calcium in each of your coronary arteries. The calcium presence is calculated to give you a "score" or number that represents your arterial calcium burden. Remember that arterial calcification is an active process mediated by bone-building cells. That causes calcium to accumulate within plaque in a consistent ratio, occupying about 20 percent of plaque volume. For that reason, the amount of arterial calcium detected with a heart scan reflects the buildup of atherosclerotic plaque. Vitamin K₂ deficiency isn't the only factor contributing to heart disease, but undercarboxylated MGP (matrix gla protein) levels, a sure sign of K₂ deficiency, do correlate to the severity of arterial calcification and the CAC score.[7] The greater the degree of K₂ deficiency, the higher the calcium score.

A high CAC score on electron beam computed tomography—a precise type of CT, discussed further below—is a better predictor of mortality than is age.[8] That means you are only as old as your arteries. For example, if you are a 60-year-old man with a low CAC score, there's a good chance you'll live to a ripe old age. On the flipside, if you are a 45-year-old man with a heavy calcium plaque burden, you are more likely to be one of the unfortunate souls who suffer a massive heart attack at age 50—if you don't take action to prevent it. Sudden death from heart attack is much more highly correlated with arterial calcification than with cholesterol.[9]

The big advantage of CAC scoring is that it quantifies your risk of heart attack. A low score means low risk, a high score means a high risk. No other risk factor offers that kind of graded risk assessment. The significance of your CAC score will depend on the scoring system used by the center that does your scan. The most widely used scoring system is Agatston scoring. Below is a typical reference table for interpreting CAC scores using Agatston scoring.

Interpreting CAC scores using Agatston scoring

CAC score	Artery blockage	Risk of heart attack in next five years	Recommendation
0–10	5% probability of significant obstruction	Very low to low	No further workup in a patient showing no symptoms
11–100	Less than 20% chance that significant obstruction is present	Moderate	Based on number of additional risk factors present
101–400	High likelihood of moderate non-obstructive coronary artery disease	Moderately high	Based on number of additional risk factors present Exercise stress testing should be considered
> 400	High likelihood of at least one obstructed coronary artery	High	Based on number of additional risk factors present Exercise stress testing should be considered

You can't get an accurate heart scan using any old CT device. Most hospital scanners that are fine for imaging stationary organs, like the brain, are not fast enough to image the beating heart. The heart appears as a blur on standard CT scans, making it impossible to precisely quantify calcium burden. A heart scanner is a precision instrument that uses electron beam computed tomography (EBT) or the even faster, more recent multi-detector CT (MDCT).

A coronary calcium scan takes about 10 to 15 minutes in total, though the actual scanning takes only a few seconds. The CT scanner is a large machine with a hollow, circular tunnel in the center. You lie on your back on a table that slides into the tunnel. If you feel anxious in enclosed spaces, you may need to take medicine to stay calm, but this isn't a problem for most people because your head remains outside the opening in the machine. During the test, the scanner makes clicking and whirring sounds as it takes pictures. It causes no discomfort, but the exam room may be chilly to keep the machine working properly.

Imaging centers offering heart scans are popping up across Canada. Your provincial health care plan may cover part of the test fee, but you will have to pay out of pocket for the rest, which may be as much as $2,000. If you live a reasonable driving distance from the border, you are probably close to an American imaging center that offers the latest heart-scanning technology for as little as US$300. Do your homework and make sure the center you choose is using EBT or MDCT technology for heart scans. You can't just show up, either; you need a doctor's referral. Many centers have online referral and requisition forms that you can print and bring to your doctor. Test results will be sent to your physician and you might leave with a copy in hand, as well.

CAC scoring is not a crystal ball. Although it can tell you how much atherosclerotic plaque is clogging your arteries, it can't tell you the location of that plaque or how severe a blockage is at any particular point. Where CAC scoring really shines is in pinpointing the risk of a cardiac event among the large group of people classified as intermediate risk according to traditional guidelines. The advantage to patients and physicians is that a CAC score can improve the accuracy of predicted risk among the patients in whom clinical decision making is most uncertain, the medium-risk people with no apparent symptoms.[10]

For example, if you're an overweight male smoker with high cholesterol and a family history of heart disease, there's a high chance that you will eventually become a heart attack victim, and your CAC score probably won't tell you anything you haven't already been ignoring. Similarly, if you are a nonsmoking, healthy weight, premenopausal female, it's unlikely you will suffer a heart attack in the foreseeable future and unlikely that you will have a surprisingly elevated CAC score. But what if you are a fit 45-year-old male with normal cholesterol and an extensive family history of heart disease? Or a postmenopausal woman with slightly elevated cholesterol? The addition of CAC scoring to the conventional prediction model based on traditional risk factors (listed below) significantly improves the classification of risk. It moves most people out of the gray zone and into a black or white category.[11] Fifty to 75 percent of intermediate-risk patients are reclassified by CAC into more accurate heart attack risk categories, leaving fewer patients in the truly "maybe, maybe-not" category.[12]

Risk Factors for Heart Disease

- overweight
- high blood pressure
- LDL cholesterol level above 2.6 mmol/L (millimoles per liter)
- HDL cholesterol level below 1.0 mmol/L
- cigarette smoking
- male over 45
- female 55 or postmenopausal
- diabetes or prediabetes
- family history of coronary heart disease

How will boosting your K_2 intake affect your calcium score? Vitamin K_2 deficiency, as measured by undercarboxylated MGP, increases arterial calcification and CAC scores. In animal studies, K_2 supplementation dramatically reduces arterial calcium burden in only six weeks. Human clinical trials got underway in 2011, and we can anticipate preliminary results within a year or two.

Should You Be Tested?

Since the real value of CAC scoring lies in discerning the likelihood of heart attack among individuals without symptoms and without major heart disease risk factors, determining exactly who should get the test is very difficult. CAC scans are too expensive to recommend broad, unselected screening, though that practice carries the best chance of identifying heart disease that might otherwise be missed. Based on the fact that calcium plaque tends to creep up sometime after the age of 40 for men and 50 for women, age is the only guideline currently in use at most testing centers. How age should be combined with other risk factors remains unsettled. If you are in the age bracket in question, concerned about your heart attack risk and can afford the test, then go for it.

It's easier to say who should not have a heart scan. In the Western world, nearly everyone over the age of 20 has some plaque buildup. CAC scoring in people with low heart disease risk mostly leads to needless expense and anxiety. If you are a man younger than 40 or a woman younger than 50 with normal blood pressure, normal cholesterol and healthy body weight and you don't smoke, it's unlikely that the calcium plaque you probably have in your arteries is life threatening.

For an in-depth look at coronary artery calcium scoring, check out *Track Your Plaque,* 2nd edition (2011, iUniverse), by cardiologist William Davis. The book explains the ins and outs of measuring coronary plaque and provides a three-step plan to identify and treat the causes of your plaque. Although the first edition of the book predates the awareness of K$_2$, the updated (and online) protocol includes this critical artery-clearing vitamin.

New Tests on the Horizon

As fascinating and useful as it is to use CT machines to peer into arteries, see calcium deposits and predict heart attack risk, heart scans are expensive and require sophisticated machinery. This complex diagnostic tool may soon be replaced with a simple pinprick. Convenient blood tests that measure the artery-clearing matrix gla protein (MGP) are just around the corner, and they may supersede CAC scoring altogether.

MGP is the famous vitamin K$_2$–dependent protein produced in bone, kidney, lung, heart, cartilage and vascular smooth muscle cell tissue. It is the main inhibitor of calcification in the heart and arteries, and MGP levels are inversely correlated with the severity of CAC.[13] When levels of carboxylated MGP increase, meaning better K$_2$ status, CAC decreases. Even research that challenges the specific relationship between MGP and CAC supports the notion that MGP varies with other risk factors for coronary heart disease, making it a useful marker for cardiovascular disease risk.[14] Scientific studies already use MGP levels as a marker of K$_2$ status and heart disease risk.

If researchers are using MGP tests, why can't we easily obtain MGP testing? What's the holdup? For one, MGP can be modified in different ways after it is synthesized. In addition to vitamin K$_2$–dependent carboxylation, MGP is potentially altered by a process called phosphorylation. Because of that, diverse forms of MGP circulate in the blood, and the practical relevance of having varying amounts of each of those forms needs to be teased out. Researchers are busy clarifying the usefulness of different

MGP testing methods currently available to come up with a clinically serviceable evaluation of MGP, vitamin K_2 and heart disease risk.

◈   ◈   ◈

Testing for vitamin K_2 status is in its infancy, though one day, rapid and reliable K_2 tests will be at our fingertips. Nutrient testing has an important place in health care and wellness management, especially in a time when nutrient deficiencies are common. That being said, there are two ends to the spectrum of approaches aimed at achieving optimum health. We can attempt to identify every micronutrient required for well-being, come up with a test to evaluate its intake and use the results of those tests to guide nutrient supplementation. Or we could strive to eat the most nutrient-dense diet possible and reclaim our health with whole foods. Given our current state of health and the food production system, the best approach will lie somewhere in between for most people, but you don't have to wait for K_2 testing to be readily available to take action.

Whether or not you know your K_2 status, a wholesome diet, along with supplements as needed for backup, will maximize your chances of staying healthy for a lifetime. It's important to remember that getting adequate vitamin K_2 is but one piece, albeit the most overlooked piece, of a bigger nutritional puzzle. Menaquinone aligns itself with vitamins A and D to overcome the Calcium Paradox. Nutrition and supplements that provide all the fat-soluble vitamins, along with necessary minerals and water-soluble vitamins, will best meet our needs. The last chapters of this story put vitamin K_2 into the bigger nutritional picture.

Vitamins K$_2$, A and D: Better Together

Things should be made as simple as possible, but no simpler.

—Albert Einstein

S ince you have already read about the amazing benefits of vitamin K$_2$, you might be tempted to skip the rest of the book and rush out to the store to buy a menaquinone supplement or, though much less likely, some natto. But wait, there's more to the story. Below, I'll to give you the quick and dirty explanation of the essential relationship between the three fat-soluble vitamins, A, D and K$_2$. Once you have read it, you may then rush out to the store, if you wish. If, however, you find fat-soluble nutrients simply fascinating, as I do, then please enjoy the rest of the chapter to get the whole story. You will learn practical information about vitamins A, D, and E and about beta-carotene, and we'll bust some myths along the way.

But first, here's the skinny on fat-soluble vitamins: vitamins A and D are required for the production of vitamin K$_2$–dependent proteins, like osteocalcin and MGP (matrix gla protein), which have enormous *potential* health benefits. Vitamin K$_2$ activates those proteins, allowing them to realize their potential and do their job in moving calcium around the body. Without adequate amounts of vitamins A and D, K$_2$-dependent proteins are not made, so K$_2$ is useless. Without adequate amounts of K$_2$, the K$_2$-dependent proteins produced by the influence of A and D cannot be activated and so remain useless. Working together, however, vitamins A, D and K$_2$ are a mighty fat-soluble triumvirate.

A, D and K$_2$: A Balancing Act

Vitamins A, D and K$_2$ are intricately interrelated in complex ways that modern science doesn't yet entirely understand. That being said, the fundamental importance of these nutrients as the foundation of good health is so critical, yet so underappreciated, that this book would be incomplete without a chapter devoted to their special relationship.

As much as "Vitamin K_2 is the ultimate supernutrient!" makes a great headline, the fact is that the benefits of K_2 are dependent on vitamins A and D—and, for many aspects of health, the opposite is also true.

The fat-soluble vitamins, A, D and K_2, are profoundly different from other nutrients. The biological role of most nutritional compounds, such as minerals and water-soluble vitamins, is to act as cofactors. A cofactor is a molecule that binds to a protein and is required for that protein's biological activity. Almost every metabolic process in the body happens because of a protein, usually an enzyme. Minerals and water-soluble vitamins are the helper molecules that allow our body's proteins to function, which is why these nutrients are so important for health.

Vitamins A and D do not act as cofactors in the way other vitamins do. The roles of A and D are more fundamental: they regulate the activity of genes that cause cells to produce the proteins to which the minerals and water-soluble vitamins will bind. Vitamin K_2—although it isn't known to affect genetic activity like A and D do—will also activate proteins, allowing them to bind calcium to do their job. It is for this reason that Dr. Weston Price referred to A, D and X (K_2) as "activators." *These fat-soluble vitamins are required so our body can make use of all other nutrients.* They are truly the foundation of health.

The relationship between A, D and K_2 is well illustrated by an image I use when lecturing about these nutrients. It's of the Three Viro Brothers, a turn-of-the-19th-century acrobatic ensemble. Each brother represents a vitamin. Vitamins A and D are at the base of the formation. Specifically, vitamin D is the brother on the left who is smiling at the camera. Everybody can see D; he is well known, popular, looking right at the audience. Similarly, vitamin D is currently the darling of the nutritional world. Good news about this nutrient is published every day,

and it all gets top billing in the media. These days, everyone knows they
need vitamin D.

3 Viro Bros

Vitamin A is the brother with his head between D's legs. His face is
obscured and his role in this trio is not as obvious as that of his brother,
D. He even seems kind of weird and scary. Likewise, the health benefits
of vitamin A are quite misunderstood and underappreciated. It has even
been unjustly maligned for being toxic. But make no mistake, without
A this pyramid would topple. Even with the benefits of K₂, the current
trend toward supplementing higher and higher doses of D will lead to
an unhealthy balance without vitamin A to maintain the equilibrium.

Above it all is K_2, the brother on top. It's obvious that without A and D, K_2 would fall flat on his face. Indeed, menaquinone would be useless without vitamins A and D to regulate the production of K_2-dependent proteins. On the other hand, you could argue that A and D don't really need K_2 up there; they could have their own little balancing act. True, many of us have gotten by with little or no K_2 in our diets for years and we're still standing. However, this stupendous feat realizes its awesome potential only with K_2 perched on top. Vitamins A and D collaborate to prop up vitamin K_2, and we fully benefit from vitamins A and D only when we have K_2 to complete the act and achieve optimal health.

In addition to their special relationship with vitamin K_2, vitamins A and D have another unique feature that distinguishes them from other nutrients: they are obtained only from animal foods. Although this might be bad news for vegans, it does explain why A and D cross the boundaries of the terms "hormone," "prehormone" and "vitamin" (discussed further below). To fully appreciate the food sources and health benefits of vitamin K_2's balancing partners, let's turn the spotlight on the information—and misinformation—about these nutritional elements.

Understanding Vitamin A

True or false: carrots are an excellent source of vitamin A.

The answer is "false." In fact, carrots don't contain any vitamin A at all. If this surprises or confuses you, it's because misleading information about vitamin A is everywhere. Retinol is one of the most important yet least understood nutrients. It has garnered a very much undeserved reputation for being toxic, which may be causing its intake to be neglected by those who need it most. A deficiency of this nutrient is much more likely than an overload, given that our consumption of vitamin A–rich foods has declined precipitously over the last century. Let's take a look at the health benefits, deficiency symptoms and food sources of vitamin A, and explore the truth about retinol toxicity.

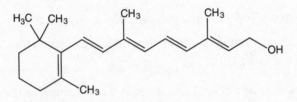

Molecular structure of retinol

Health Benefits of Vitamin A

Vision

Vitamin A has been known for at least 3,500 years as a factor in healthy vision. As far back as the ancient Egyptians, many cultures have recognized that eating liver (an excellent source of vitamin A) would restore vision in people with night blindness.[1] The scientific name for vitamin A is retinol, reflecting its presence in the retina of the eye. One of the first signs of vitamin A deficiency is a reduced ability to see in dim light.

Vitamin A has a specific, highly complex function in vision under low-light conditions. The body converts dietary retinol into a form of vitamin A called retinal. Retinal binds with opsin, a protein in the retina, to create the light-sensitive pigment rhodopsin, also known as "visual purple." Rhodopsin resides in the part of the eye responsible for low-light and peripheral vision. Upon exposure to light, the visual pigment breaks apart, releasing energy signals that the brain interprets as a picture.

Seeing in the dark

An early modern vitamin A researcher was moved to make the following poetic remark on this topic: "It may be an inspiring thought... that Man's knowledge of the existence of the stars and the vast universe which appears in the heavens each night, comes in the first place from the stimulation by light rays of delicately poised molecules of vitamin A."[2] The next time you find yourself gazing at the stars, you can thank vitamin A.

Severe vitamin A deficiency leads to blindness, even in daylight, a condition called xerophthalmia. In his pivotal work, *Nutrition and Physical Degeneration*, Dr. Weston Price tells the story of a prospector who, while crossing a high plateau in the Rocky Mountains, went blind with xerophthalmia due to lack of vitamin A. As he sat weeping with despair and, quite literally, blinding pain, he was discovered by a Native. The Native led the blind man by the hand to a stream, where the Native caught a fish and instructed the prospector to eat the flesh of the head and the tissues near the eyes, as well as the eyes themselves. Within a few hours the prospector's pain had subsided, and within two days his eyes were back to normal.[3] Despite that the men did not share a common language, the Native recognized the blind man's plight and knew how to quickly obtain a food that would cure the condition—a food rich in vitamin A.

Skin and Epithelium

Retinol is essential for maintaining the healthy integrity of tissues that line the outside and the inside of the body. The skin is one of the first organs to show signs of vitamin A deficiency. The epithelium is the delicate skin-like tissue that lines the mouth, nose, throat, eyes, stomach, digestive tract, bladder, urinary tract, vagina and almost every bodily organ. These cells act as important barriers to invading microorganisms. Without adequate retinol, the structure and function of these tissues is compromised.

Immune System

Is taking cod liver oil a part of your cold and flu prevention routine? It should be. Vitamin A has long been known as the *anti-infective vitamin* because it plays an essential role in protecting the body from infection. Retinol seems to boost immunity in several ways, including the maintenance of healthy skin and respiratory tract tissue and optimal production of antibodies and white blood cells in response to foreign bacteria and viruses. Subclinical vitamin A deficiency (low vitamin A levels without any apparent symptoms) is associated with an increased risk of a wide range of infections, from colds and flus to HIV.

Bones and Teeth

Vitamin A is required for normal development, growth and maintenance of the skeleton throughout life. Specifically, it increases the number and activity of osteoclast cells, which break down bone tissue. Although this sounds like it would be bad for bone health, it is actually necessary for the ongoing process of skeletal maintenance called bone remodeling. In this way, old or weakened bone tissue is removed to make room for new, stronger tissue. This action is critical to fracture repair and retaining bone density. Vitamin A also plays a role in stimulating osteoblasts (the bone-building cells) to secrete proteins that are required for bone mineralization, including the K$_2$-dependent osteocalcin.[4]

Fetal Development

Retinol is crucial for the healthy development of almost every part of the growing fetus, from the central nervous system and limbs in the first few weeks after conception to the lungs in the last few weeks before birth. A deficiency of vitamin A can have serious health consequences for the baby. For example, fetal alcohol syndrome, a well-defined group of birth defects, is thought to be caused by an alcohol-induced vitamin A deficiency; the mother's liver is so busy detoxifying ethanol it cannot convert retinol into the active form of retinoic acid, and the growing fetus suffers. Even mild vitamin A deficiency during pregnancy can have a lifelong impact on health.

Fertility and Pregnancy

Vitamin A is essential for reproductive functions both in men and women. In men, retinol plays a key role in sperm production.[5] In women, it is necessary for the production of estrogen, progesterone and a number of other hormones directly related to reproduction. Women with mild vitamin A deficiency are at a higher risk for placental abruption, a dangerous, premature separation of the placenta from the uterus wall, and inadequate milk supply.[6] In developing nations, UNICEF and the World Health Organization recommend high-dose retinol supplementation for women in the immediate postpartum period, coupled with exclusive breast-feeding, so that all infants receive the necessary immune-boosting protection of vitamin A in the first six months of life.[7]

Cancer Prevention

Throughout life, vitamin A works at the genetic level to ensure tissue differentiation, a process by which individual cells mature into a specific and well-defined type of tissue. Normal, healthy cells are well differentiated; they show obvious structural features that are characteristic of their tissue type. Cancerous cells are poorly differentiated; they look more like indistinct blobs. Vitamin A inhibits tumor development, especially those of epithelial origin, such as colon and cervical cancer, by promoting healthy tissue differentiation. Low plasma retinol so strongly predicts poor prognosis in postmenopausal breast cancer patients that experts now recommend retinol levels be evaluated as part of the prognostic workup.[8]

Vitamin A Deficiency

Severe vitamin A deficiency is a major public health concern in developing nations. It is one of the most serious and widespread nutritional disorders contributing to disease and death, especially among children. UNICEF, the World Health Organization and several other health and aid organizations have massive, ongoing retinol supplementation campaigns in almost 40 countries to help the hundreds of millions of adults and children affected by this nutrient deficiency. Since supplying a daily dose of retinol can be logistically challenging in developing countries, supplementation campaigns typically aim to provide a very high dose of vitamin A (100,000 international units, or IU, for children up to 12 months of age and 200,000 IU for adults and children over 1 year) every four to six months.

Inadequate vitamin A status is not only a plight for developing nations. Mild vitamin A deficiency produces symptoms that are subtle and easy to overlook but may still affect your health. Run your hand over the back of your upper arm. Do you feel dry, rough skin with many small, hard, pimple-like bumps? That's technically referred to as follicular hyperkeratosis, and it's a sign of vitamin A deficiency.[9] Other symptoms include dry skin, thinning hair, brittle nails and dry eyes. One sign of marginal vitamin A deficiency that's easily missed, since we are never far from a light switch, is poor night vision. In the elderly especially, many of these retinol deficiency warning signs may be mistaken as a normal part of aging, compounding their already high risk of nutritional deficiencies. Up to 15 percent of seniors are vitamin A deficient.[10]

Subclinical (that is, without apparent symptoms) vitamin A insufficiency has been shown to induce changes in the respiratory tract that can impair resistance to infection, particularly in children. A study of Australian children who were not obviously deficient in vitamin A but who suffered from frequent colds and flu showed that those who received around 1,500 IU of retinol supplements daily had significantly fewer respiratory infections than children who received a retinol-free placebo.[11]

One consequence of vitamin A deficiency that remains under-investigated is growth and development of children with so-called healthy diets. As vitamin A is required for protein utilization, the high-protein, low-fat diets that are currently promoted by many conventional nutrition experts risk depleting vitamin A stores. At least one fat-soluble-nutrient expert has observed that this type of nutrition results in "tall, myopic, lanky individuals with crowded teeth and poor bone structure, a kind of Ichabod Crane syndrome . . . [that] are a fixture in America."[12]

How common is mild vitamin A deficiency in the seemingly healthy populations of developed nations such as our own? It's hard to know for sure. Unlike vitamin D, blood tests are not reliable indicators of vitamin A status. Blood tests for retinol reflect only extremely high or low vitamin A intakes, and "normal" blood levels of vitamin A vary widely within groups of apparently healthy individuals. The most precise way to assess vitamin A nutritional status is with a liver biopsy, which is too invasive to test large numbers of people, so definitive data are lacking.

Dietary intake surveys are also of limited usefulness when it comes to estimating vitamin A intake because so many factors affect retinol absorption from food, such as fat composition of the meal and the amount of retinol already circulating in the blood or stored in the liver. Despite the lack of a useful screening test for vitamin A status, there is enough evidence for experts to conclude that retinol deficiency is "probably under-recognized in the United States and other developed countries who do not normally consider their citizens to be malnourished."[13]

Underlying the possibility of widespread subclinical retinol deficiency in our own "well-nourished" population is the fact that we are eating less vitamin A–rich food than ever. Weston Price discovered that the diets of healthy traditional people contained at least *10 times* more vitamin A than the standard American diet of the 1930s. Since the regular consumption of vitamin A–rich foods has decreased even further since that time, our current intake of retinol is likely very paltry compared with the levels that kept traditional people healthy. Do you remember when a daily dose of cod liver oil and a weekly serving of liver were just a part of the family routine? I don't, but I've heard about it from my parents. That nutrition wisdom has now fallen by the wayside and with it our last rich sources of dietary vitamin A. The table below lists the vitamin A content of some foods.

Vitamin A content of selected foods

Food	International units
Cow's liver, cooked, 3 oz	27,185
Chicken liver, cooked, 3 oz	12,325
Whole milk, 3.25% milk fat, 1 cup	249
Cheddar cheese, 1 oz	284
Whole egg, 1 medium	280

Source: U.S. Department of Agriculture, Agricultural Research Service. 2004. USDA National Nutrient Database for Standard Reference, Release 17.

Anemia and vitamin A deficiency

In both adults and children, vitamin A deficiency produces a mild anemia that can be corrected with vitamin A supplementation.[14] Many people who struggle with anemia focus only on iron intake to correct the problem. Ensuring adequate retinol intake addresses anemia from another angle.

Is Vitamin A Toxic?

Unlike water-soluble nutrients, which tend to be excreted freely in urine, fat-soluble vitamins have the potential to accumulate in the body's tissues, possibly leading to toxic effects. In reality, when it comes to retinol, this happens rarely and only in individuals who supplement with extremely high doses for months on end. Unfortunately, reports of vitamin A toxicity have been grossly exaggerated, creating an unwarranted fear of this vital nutrient.

Toxicity research shows that injury from vitamin A results from excessively high and prolonged intakes of retinol supplements and not from foods. In adults taking a whopping 100,000 IU every day, negative effects manifest in as early as six months, though some individuals do not show toxic effects until they have been on this excessive dose for several years. Symptoms are generally reversible once the excessive intake has stopped.[15]

Unfortunately, the fear-mongering about the toxicity of vitamin A has not been focused on abusively high intake levels; rather, it has targeted more reasonable retinol intakes. Specifically, so the story goes, anything above the recommended daily allowance of 3,000 IU for men and 2,300 IU for women (4,300 IU for pregnant women) is potentially dangerous. There is simply no evidence to support this claim. No adverse effects have been observed in long-term studies in people consuming twice the recommended daily allowance of vitamin A.[16] Further studies using up to 25,000 IU daily for periods ranging between 2 and 12 years showed no liver damage or other toxic effects.[17] (An exception to the general vitamin-A-isn't-so-toxic rule is for those who abuse alcohol or have liver disease, because the liver is involved in retinol metabolism. If you have a history of excessive alcohol intake or impaired liver function, consult a nutritionally oriented physician before taking vitamin A supplements.)

What are the toxic effects of excessive, long-term vitamin A supplementation? Symptoms include loss of appetite; dry, itchy skin; hair loss; headache; bone thickening and liver damage. You may notice that, perplexingly, several of the symptoms of vitamin A excess are the same as those of vitamin A deficiency. That is because *many of the symptoms of vitamin A toxicity are actually caused by an induced deficiency of vitamins D and K2 and vice versa.* To clarify, when you take more of one fat-soluble vitamin, you create a greater need for the others. If those others are lacking, toxicity symptoms result. We'll delve into this concept in our discussion of vitamin D below.

Does Vitamin A Antagonize Vitamin D?

Ironically, the rising popularity of vitamin D seems to have cast an unflattering shadow on its balancing partner, vitamin A. Even the California-based Vitamin D Council's official, though somewhat cryptic, recommendation is to limit vitamin A intake to "tiny amounts." That's because retinol is said to oppose or antagonize the health benefits of vitamin D. For example, one of the health roles of vitamin D is to activate osteoblasts, the bone-building cells, to increase bone density. As you just learned, a physiological role of vitamin A is to promote bone breakdown so new bone can be laid down. Although these processes seem to oppose one another, they are both necessary to maintain bone health. It is more accurate to describe the actions of vitamins D and A as complementary, not antagonistic.

In the words of one fat-soluble-vitamin expert, "That vitamin A antagonizes some of the effects of vitamin D is no more reason to avoid vitamin A than the fact that vitamin D antagonizes some of the actions of vitamin A is a reason to avoid vitamin D."[18] Indeed, when it comes to the production of K₂-dependent proteins, D and A are like a car's gas pedal and brake pedal respectively. You need both to drive. More importantly, the actions of vitamins A and D are not completely discrete or mutually exclusive. For example, the two nutrients join forces to stimulate osteocalcin production. Although vitamin D on its own will increase osteocalcin production, together vitamins A and D have what scientists describe as a "remarkable synergistic effect" to boost osteocalcin output.[19] If vitamin A truly antagonized vitamin D, you'd expect the former to cancel out the latter and to see no benefit when given simultaneously. Instead, they cooperate for a bone-building bonus that's greater than the sum of its parts.

Does Vitamin A Cause Osteoporosis?

Another popular rumor about vitamin A is that it contributes to the development of osteoporosis. This misconception is based on two types of evidence, intervention studies and population-based research.

Knowing what you now know about the effect of vitamin A on osteoclasts, it's not hard to predict that administering high amounts of vitamin A without vitamin D would be bad for bone density. And therein lies the major shortcoming of human studies of fat-soluble-nutrient toxicity: they examine each vitamin in isolation. Animal study results, on the other hand, illustrate an all-important principle. Administer enough of either vitamin A or D on its own and you will eventually see toxic effects. When you supplement with *both A and D together, you will never see symptoms of toxicity of either vitamin*, no matter how high the intake of A and D.[20]

Epidemiological evidence does show that countries with the highest vitamin A intakes—Scandinavian nations whose diets favor retinol-rich foods like liverwurst—tend to have the highest bone fracture rates. However, these studies rarely take into account vitamin D levels, a critical factor in fracture risk, even though the very same countries are known to have low vitamin D levels because of their long winters. One of the few researchers to actually consider the vitamin A-D balance when investigating fracture rates concluded, "The high intake of vitamin A in Scandinavia may aggravate further the effect of [low levels of vitamin] D on calcium absorption."[21]

Does Vitamin A Cause Birth Defects?

One area of special concern is the increasingly popular recommendation that pregnant and lactating women limit their intake of vitamin A to very minimal levels or avoid it altogether. The rationale for this misguided advice is that high-dose vitamin A has been shown to cause birth defects, so extreme caution should be employed with any retinol intake during pregnancy.

The potentially birth defect–causing effects of vitamin A were not discovered with normal dietary intake, or even with vitamin supplements. The medical community noticed vitamin A–associated birth defects in the 1980s when pregnant women were prescribed isotretinoin, an acne medication sold under various brand names, including Accutane. Isotretinoin, a retinol derivative also known as 13-cis-retinoic acid, does occur naturally in the body in small quantities. The massive amounts of the synthetic form of this compound in acne medication induced a characteristic pattern of pronounced deformities in infants who were exposed to it in the womb. This established the potential of retinol to cause birth defects in humans.

Of course, fetal development is not an area where we want to take any chances, but data from several large studies show that prenatal supplements containing vitamin A do not increase the risk of any of the malformations that have been linked to isotretinoin.[22] Some inconclusive evidence suggests a small increased risk of birth defects in the offspring of women taking high-dose vitamin A (over 25,000 IU) on top of what is contained in a prenatal multivitamin, but still nothing close to what is seen with isotretinoin. Despite this, many health authorities (including those who should know better, like the U.S. Food and Drug Administration) have thrown the baby out with the bathwater by adopting a zero-tolerance policy on vitamin A supplementation. It suggests that women of reproductive age attempt to meet their vitamin A requirements by relying only on foods containing beta-carotene, a very risky recommendation, as you'll see shortly.

Women of childbearing age are especially vulnerable to vitamin A deficiency because of the extra requirement of this nutrient during pregnancy and lactation. Vitamin A is also vitally important for the growing fetus. Even mild vitamin A deficiency can cause defects in organ development that are not recognized at birth but that lead to long-term health consequences.[23] As such, it is very shortsighted to advise expectant mothers to limit vitamin A intake and avoid vitamin A supplements, since there are more documented risks associated with inadequate vitamin A intake than with excess vitamin A intake.

Pregnant women, and their growing babies, need plenty of all the fat-soluble vitamins. Expectant moms should regularly enjoy foods from the list of those high in vitamin A on page 193, and balance that with a diet rich in vitamins D and K₂. Women who are planning a pregnancy or who might conceive and do not eat vitamin A–rich foods will benefit from taking a prenatal multivitamin containing between 5,000 and 8,000 IU of natural-source vitamin A.

Beta-Carotene Is Not Vitamin A

The common misconception about carrots containing vitamin A stems from the fact that they, along with other orange vegetables and leafy greens, are a source of beta-carotene. Beta-carotene is a potential precursor to vitamin A. Retinol is sometimes referred to as "preformed vitamin A" to distinguish it from beta-carotene, which needs to be converted in the body. Unfortunately, the vast majority of books and otherwise reliable websites on nutrition perpetuate the inaccurate notion that vitamin A and beta-carotene are virtually one and the same. Reinforcing this confusion, U.S. and Canadian regulations allow food packagers to label beta-carotene content as vitamin A, giving consumers the erroneous impression that a single serving of carrots can provide 110 percent of the daily requirement of retinol. In truth, it is almost impossible to predict how much vitamin A you are getting from a serving of vegetables, and some research shows you might not be getting any at all.

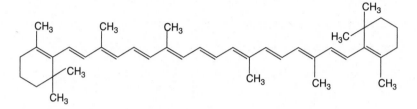

Molecular structure of beta-carotene

In its molecular structure, beta-carotene looks like two molecules of retinol stuck together, end to end. In theory, once beta-carotene is absorbed, the body could just cleave each molecule in half to produce two molecules of vitamin A. In reality, it doesn't happen like that— not even close. Calculations from clinical trials of the actual rate of conversion range from 6 to 1 (meaning it takes six molecules of beta-carotene to yield one molecule of retinol) up to 48 to 1.[24] The amount of vitamin A listed on the package labels or nutrient content tables of plant-based foods is based on optimal absorption and conversion of the beta-carotene in those foods. Strong evidence suggests these published values are seriously overestimated.[25]

It is common to hear that the body readily converts beta-carotene into retinol and, somehow, converts only as much as is needed, thus avoiding possible toxicity and therefore making beta-carotene superior to preformed vitamin A. Even nutrition experts perpetuate the myth that, since the body can convert beta-carotene to vitamin A, the body's requirements for this vitamin can be met entirely by beta-carotene. In reality, populations that rely on beta-carotene for their vitamin A intake are at risk of vitamin A deficiency.

Why isn't beta-carotene readily converted to retinol? First, absorption is an issue. The absorption rate of beta-carotene is only 20 percent to 50 percent that of retinol, and it is well established that the more beta-carotene a food contains, the less it is absorbed and converted to vitamin A.[26] Second, many common health conditions interfere with the conversion of beta-carotene to vitamin A,[27] including:

- diabetes
- low thyroid function
- low fat intake (dietary fat is necessary for beta-carotene conversion)
- nutritional deficiencies of zinc or protein
- celiac disease
- absent gallbladder
- being a child or infant
- intestinal roundworms
- tropical sprue

The last two conditions listed are common in developing countries where vitamin A deficiency is rampant and where genetic modification of rice to include beta-carotene (so called "golden rice") is being considered as a replacement to current retinol supplementation programs.

Even in healthy people, evidence suggests that the rate of absorption and conversion of beta-carotene is much lower than is commonly believed, from as little as zero to a maximum of 50 percent.[28] A study of breast-feeding mothers in Indonesia showed that the intake of dark green leafy vegetables providing enough beta-carotene and dietary fat to yield, in theory, three times the recommended amount of vitamin A failed to improve vitamin A status and instead left the women vitamin A deficient.[29] This should serve as cautionary information for experts who advocate that pregnant and breast-feeding women rely on beta-carotene to meet their need for vitamin A.

Beta-carotene certainly does have its health benefits. It is an antioxidant that seems to play a role in heart health and cancer prevention. Although some beta-carotene, under optimal conditions, can be converted to retinol, its merits are best considered apart from its role as a potential precursor to retinol—any conversion to vitamin A is a bonus.

The Relationship between Vitamins A and K_2

Vitamin A plays a valuable role in managing the body's need for vitamin K_2. On a molecular level, it is precisely this misunderstood role that gave vitamin A an undeserved bad rap. Vitamin D stimulates the production of vitamin K_2–dependent gla proteins, thereby increasing the body's demand for vitamin K_2 and the potential to benefit from K_2. That makes vitamin D a superstar because the more vitamin K_2–dependent proteins you make, the more calcium you can direct into bones and away from arteries, if you have the K_2 to activate those proteins. So vitamin D looks good.

Here's where things get tricky. Working together, A and D synergistically improve osteocalcin production. On its own, however, vitamin A limits the production of MGP. This sounds detrimental to heart health— and in large amounts it would be—but it effectively minimizes the body's requirements for K_2. Vitamin A has a K_2-sparing action; having adequate amounts of retinol reduces the demand for K_2, allowing your body to get by on less menaquinone.[30] When K_2 is scarce, vitamin A does damage control. Of course, if this is taken to an extreme with long-term high-dose vitamin A supplementation paired with a lack of vitamin D, you will eventually see problems like lower bone density because you have spared K₂ too much. As with the Three Viro Brothers, it's all about balance.

Vitamins A and K₂ have an additional buddy system in which retinol complements the action of vitamin D. It is yet another realm in which the effects of vitamin A are misconstrued. Remember the seasonal variations in calcium excretion that parallel changes in arterial calcifications? Calcium plaque diminishes in late summer, while bone density is retained. Where is the extra calcium going? Down the toilet.

There is a seasonal variation in urinary calcium excretion (yes, it's been studied), and it is not what you would expect. Given that bone density loss occurs almost exclusively in winter, you might predict urinary calcium loss to be highest at this time. Instead, calcium excretion is minimal in winter months when bone density is diminishing—and when arterial calcifications are growing.[31] The calcium being lost from bone is not leaving the body; it is being transferred to plaque.

In the northern hemisphere, urinary calcium peaks in August, but bone density remains stable so that lost calcium is not being leached from our skeleton. Arterial calcifications, on the other hand, diminish in the same month. What prompts the body to rid itself of calcium at this time of year? It just so happens that retinol levels also follow an annual cycle, peaking in summer.[32] Curiously, although vitamin A intake is generally constant year-round, blood levels of retinol and its carrier protein are elevated in the summer. Retinol is famous for causing urinary calcium loss, an effect that has long been maligned as promoting osteoporosis. Looking at annual trends, we see that when retinol levels and calcium output are highest, bone density isn't affected.

Instead, arterial calcifications shrink. Vitamin A is triggering the body to release the calcium that K_2 has liberated from the arteries.

Since retinol regulates the production of K_2-dependent MGP, it seems to be at odds with the action of menaquinone. And yet, K_2 content in grass-fed foods varies with the content of retinol. Why would K_2 pair up with its apparent nemesis? Vitamin A promotes urinary excretion of calcium. When K_2 removes calcium from arterial plaque, vitamin A disposes of it. Vitamin D promotes calcium absorption so K_2 can guide it into bones and teeth. Vitamin A chaperones calcium out of the body when K_2 has extracted it from soft tissue. This is the calcium cycle of life.

So exactly how much A, D and K_2 do we need to keep everything in balance and keep calcium in its proper place? Since vitamin K_2 does not act like a hormone and stimulate protein production, it has no toxic effects, as you know. For that reason, K_2 isn't a limiting factor—it will activate as much osteocalcin and MGP as it finds. It's vitamins A and D that are the limiting factors. Since they are also the nutrients with potentially toxic effects, we need to determine our relative requirements of these vitamins. How much A do we need to optimize the benefits of and prevent potential toxicity of D and vice versa?

There is no established optimal ratio for vitamins A and D, although several have been suggested based on educated guesswork. The most intelligent analysis of this question proposes that there is not an optimal ratio of vitamins A and D per se. Rather than A and D interacting by a ratio model, they interact by a switch model whereby a minimum amount of D switches off the potentially toxic effects of A and a minimum amount of A switches off the toxicity of D.[33] As long as you have some of each, then, not only are you protected but you stand to reap the biggest benefits. Vitamins A and D will work synergistically to maximize production of osteocalcin and MGP. Be sure you have plenty of menaquinone available to activate all those proteins so they don't go to waste. A diet rich in all the fat-soluble vitamins will accomplish that.

A switch model makes good sense in light of the fact that traditional diets around the world would have had a varying relative intake of these nutrients. In regions where both seafood and sunshine are staples, vitamin D intake might be relatively high compared with A. The opposite might be true in areas where the organs of land-dwelling creatures were a mainstay. Mother Nature couldn't afford to be picky about a particular ratio of A and D; she'd be happy as long as you were getting minimum amounts of each and hopefully lots of both. This is where the wisdom of obtaining fat-soluble vitamins from food shines through. Although supplements will be necessary to bridge the gap while dietary intake of A, D, K₂ and E catches up to demand, a thoughtfully diverse diet will keep it in balance.

Understanding Vitamin D

Unless you have been living in a cave for the past few years—which causes vitamin D deficiency, by the way—you have probably heard some good news about the sunshine vitamin. Low vitamin D levels have been linked to cancer, diabetes, high blood pressure, osteoporosis, irritable bowel disease and numerous other health conditions. Vitamin D deficiency has even been found to be an independent predictor of all-cause mortality.[34] In other words, regardless of any other health risk factor, a lack of vitamin D will increase your risk of death. That's a pretty powerful statement about a vitamin.

Although we discovered vitamin D in the 1920s, 99 percent of what we know about this nutrient we learned in the last decade, and we've only just scratched the surface. As with vitamin A, there are vitamin D receptors in most, if not all, cells of the body. This points to a fundamental importance of both these vitamins for our well-being. While a deficiency of vitamin D is linked to a wide range of illnesses, vitamin D's specific mechanism of action in those conditions isn't as clearly defined as it is with many vitamin A–deficiency problems. Given that a whole book could be devoted to vitamin D, and several already exist, I'll focus on the effects of vitamin D as they pertain or likely pertain to vitamin K₂.

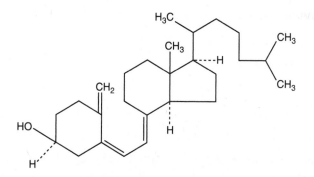

Molecular structre of vitamin D₃

Health Benefits of Vitamin D

Bone Health

Vitamin D has been most famous as a key player in bone health since it was used to eradicate rickets in the 1930s. Among other things, vitamin D promotes calcium absorption from the intestines, allows for proper function of parathyroid hormone to maintain blood calcium levels and increases the number of osteoclasts, the bone-breakdown cells. That last action may be surprising, since we think of vitamin D as a nutrient that builds bones (and it does), but it goes to show that the actions of vitamins D and A are not entirely black-and-white.

Taking a minimum of 800 IU of vitamin D daily, along with calcium, will completely offset winter bone loss and lessen the risk of osteoporotic hip fractures. Interestingly, vitamin D reduces the occurrence of fractures within the first year of taking it, well before bone density is boosted enough to have an effect on fractures. This is probably because vitamin D enhances muscle strength and balance, leading to fewer falls that can cause fractures.[35] As falls can have major and minor health consequences other than fractures, it's worth supplementing with vitamin D for this perk alone.

Cancer Prevention

Population-based studies show that higher blood levels of vitamin D and/or sun exposure are associated with lower rates of every major form of cancer, including breast, ovarian, prostate, colon, lung, non-Hodgkin's lymphoma and others. As with vitamins A and K₂ deficiency, a lack of vitamin D predicts a worse prognosis for some types of cancer, notably breast cancer. The first randomized, controlled intervention trial to supplement with vitamin D, the results of which were published in 2007, showed an astounding 60 percent reduction in all types of cancer in the women taking calcium and 1,000 IU of vitamin D daily versus women taking calcium and a vitamin D–free placebo.[36]

Obesity Prevention

In both children and adults, body fat accumulates as blood levels of vitamin D fall. That's the opposite you'd expect for a fat-soluble vitamin, hinting at some causal relationship between vitamin D deficiency and obesity. Although, like adults, most children are vitamin D deficient, the severity of deficiency correlates with the severity of obesity.[37] Along with obesity comes type 2 diabetes, and research in adults shows that vitamin D deficiency also increases that risk. Although researchers are still officially in the dark about the connection between the sunshine vitamin and diabetes, some scientists—and astute readers of this book—are already suspicious of one possible connection. Vitamin D stimulates production of osteocalcin, which in turn improves insulin sensitivity when activated by K_2. Less vitamin D means less osteocalcin to be activated and desensitized insulin response.

Several intervention trials have tested whether taking vitamin D supplements improves insulin sensitivity, with inconsistent results.[38] Vitamin D supporters claim this is because the studies conducted were not well designed. That may be so, but inconsistencies are not surprising if the real benefit of vitamin D for diabetes depends on vitamin K_2. Supplementing with vitamin D will produce more osteocalcin, which in turn will boost insulin sensitivity, if you have the K_2 to make it work. Vitamin D won't improve insulin sensitivity as effectively in K_2-deficient participants, making it a confounding factor. The vitamin D–diabetes connection will remain unclear until menaquinone status is taken into account.

Blood Pressure Lowering

There is a significant inverse relationship between blood levels of vitamin D and hypertension.[39] This manifests in many observable ways. For example, the further you travel from the equator, the more people you find with high blood pressure. The population prevalence of hypertension grows as you move from the sunniest toward the darkest, most vitamin D–deficient regions of the planet. Similarly, high blood pressure is more common in winter than in summer.[40] Disparity in vitamin D levels also partially accounts for a pronounced racial disparity in hypertension. Vitamin D deficiency is more common among blacks, who are also more frequently afflicted with hypertension than whites.[41]

In a perfect example of the triage theory, women with normal blood pressure who are vitamin D deficient before the age of 45 are three times more likely to suffer high blood pressure after 60 years of age.[42] Fortunately, it's not too late for vitamin D to help remedy the situation, even at an advanced age. Supplementing with at least 800 IU of vitamin D lowers blood pressure in women in their 70s.[43]

Vitamin D controls and lowers blood pressure by several mechanisms, none of which seem related to vitamin K₂. So what's the K₂ link to this vitamin D benefit? While calcification causes arterial stiffness that leads to hypertension, the relationship is indirect. To date no one has studied whether K₂ deficiency exacerbates hypertension or whether menaquinone supplementation will lower high blood pressure. However, individuals with high blood pressure are more likely to die from coronary artery disease than individuals with arterial calcification and normal blood pressure. A lack of vitamin D compounds the cardiovascular dangers of vitamin K₂ deficiency and vice versa.

Multiple Sclerosis

The occurrence and progression of multiple sclerosis (MS) follows a pattern similar to hypertension. There is striking annual variation in the number and severity of MS lesions that parallels blood levels of vitamin D within the same patients.[44] The effect has a consistent two-month lag time: when vitamin D levels increase or decrease, there is a subsequent increase or decrease in MS lesion activity about two months after. Chapter 5 details the K_2 tie-in.

Juvenile Diabetes

Type 1 diabetes is an autoimmune disease in which the body's immune system attacks and destroys the insulin-producing cells of the pancreas. Type 1 diabetes strikes children and adults suddenly, leaving them dependent on injected or pumped insulin for life, and carries the constant threat of devastating complications.

While the causes of juvenile diabetes are not yet entirely understood, both genetic factors and environmental triggers are involved. Of the environmental factors, vitamin D intake is extremely important. The use of cod liver oil during pregnancy reduces the occurrence of type 1 diabetes in the mothers' offspring.[45] A large population-based study conducted in Finland showed that vitamin D supplementation of 2,000 IU per day in infants lowers the risk of developing type 1 diabetes by a remarkable 85 percent.[46] A possible K_2 connection for juvenile diabetes remains unexplored. However, since the pancreas sequesters vitamin K_2, and menaquinone is intimately involved with type 2 diabetes, its involvement with type 1 diabetes begs to be investigated.

Immunity and Heart Disease

Vitamin D boosts immune system function in several ways, including the production of compounds called cathelicidins. These natural antimicrobial agents will destroy bacteria, fungi and viruses, helping to prevent a number of infections, including seasonal colds and flus. Dr. John Cannell, executive director of the nonprofit Vitamin D Council, asserts that doses of 2,000 IU of vitamin D taken daily for three days might produce enough cathelicidin to cure common viral respiratory infections such as influenza and the common cold.[47]

I promised to limit the discussion of vitamin D benefits to those connected to vitamin K₂ in some way. The physiological role of vitamin D in immune system function does not appear to pertain to vitamin K₂, but I'm bending my rule because it does relate to heart disease in a very unexpected way: a growing body of evidence demonstrates that heart attacks may be contagious.

Chlamydophila pneumoniae is a common cause of pneumonia, as well as pharyngitis (sore throat), bronchitis and sinusitis. This bacteria was previously called *Chlamydia pneumoniae*, but the name was changed to avoid confusion with the sexually transmitted disease known as chlamydia. In addition to causing between 5 percent and 10 percent of ordinary winter illnesses, this bug might actually give rise to atherosclerosis; indeed, the evidence is compelling. The association of *C. pneumoniae* with atherosclerosis is corroborated by the presence of the bacteria in atherosclerotic lesions in major arteries—coronary and carotid—and the near absence of it in healthy arterial tissue.[48]

Here's where the sunshine vitamin comes in. Cathelicidins, the natural antibiotic agents whose production is powerfully boosted by vitamin D, effectively kill off *C. pneumoniae*.[49] In addition to limiting the amount of MGP available to be activated by K₂, a lack of vitamin D impairs the immune system, opening the door for infection that may trigger plaque formation.

Vitamin D Deficiency

Vitamin D deficiency is endemic in the developed world. Our biology was designed by evolution for life in equatorial Africa. Sunshine was always our source of vitamin D; there was no significant amount of vitamin D in the foods that our Paleolithic ancestors would have been eating. Put another way, we naked apes didn't need vitamin D in foods, as the sun always provided it for us. Humans managed to migrate away from the cradle of life and thrive in climates with long winters, largely thanks to one particular adaptation. Limited vitamin D caused natural selection for white skin, which makes vitamin D from sunshine more readily available than dark skin, which is adapted for sun protection. In northern latitudes, dark-skinned women with vitamin D deficiency would produce fewer offspring than light-skinned mothers, whose skin could synthesize enough vitamin D for pregnancy and lactation.[50] People who lived closest to the poles, where cool climates necessitate covering up skin, thrived by relying heavily on one of the few worthwhile dietary sources of vitamin D, seafood.

Humans have not thrived, however, since we migrated indoors. With the industrial revolution came factory employment and rickets, a disease of sunshine/nutrient deficiency that plagued industrialized nations until the missing nutrient was identified. We 21st-century humans now spend 90 percent of our time indoors. During the little time we might be in the sun, we are 95 percent covered up with clothing or sunscreen. Shockingly, despite vitamin D fortification of foods, rickets is making a comeback in developed nations.[51] More insidiously, vitamin D deficiency is a player in most diseases that we have come to accept as commonplace.

Since we evolved in sunny, vitamin D–abundant conditions, our vitamin D metabolism is effectively designed to adjust for higher inputs, not lower inputs.[52] In other words, the body has safeguards to protect against the ill effects of excess vitamin D, but it can't compensate for chronic low vitamin D intake. Vitamin D insufficiency triggers the triage effect. When supplies are restricted, vitamin D metabolism is channeled only to the immediate needs of calcium-related functions. Immune system actions that would prevent juvenile diabetes, high blood pressure, MS and cancer are sacrificed. As a result, just as with conditions caused by vitamin K_2 deficiency, the accepted, "normal" prevalence of many conditions could be lessened substantially by increasing our intake of vitamin D.

An indoor lifestyle impacts dark-skinned people much more severely than their light-skinned neighbors. Increased skin pigmentation can reduce cutaneous vitamin D_3 production as much as 99.9 percent, which may explain, at least in part, the higher prevalence of vitamin D deficiency in the populations of African-Caribbean descent.[53] For people whose skin tone is adapted to withstand maximum sun exposure, intentional or unintentional sun avoidance is potentially deadly. Black-white differences in blood pressure, due to relatively greater vitamin D deficiency, account for thousands of excess deaths from heart attack and strokes among black individuals each year.

Vitamin D Toxicity

With all the good news about vitamin D lately, we are hearing less about its potential toxicity. That's odd, since vitamin D has as much toxic potential as vitamin A—although the potential for each has been highly exaggerated. Fear of vitamin D toxicity and a lack of knowledge about the benefits of vitamin D have kept the recommended daily intake at a barely sufficient minimum for years. Even the current "upper tolerable" limit of 2,000 IU of vitamin D daily isn't enough to raise blood levels to desirable, therapeutic amounts for most people, never mind being harmful. Sun exposure can provide up to 10,000 IU per day, and similar oral intakes are harmless.[54]

The toxic symptoms of excess vitamin D intake can be summarized with two words: inappropriate calcification. Too much vitamin D leads to excess calcium in the blood and urine, kidney stones, soft tissue calcification and softening of bones, called osteomalacia. This sounds like the Calcium Paradox, doesn't it? That's because increasing vitamin D amplifies the body's need for K_2. Vitamin D toxicity is an acute, induced deficiency of K_2 and A. If the amount of K_2-dependent proteins stimulated by taking D exceeds the supply of K_2, the Calcium Paradox sets in. Vitamin D accelerates the calcifications seen with vitamin K_2 deficiency.[55]

Two forms of vitamin D are available in supplements, D_2 and D_3, with a big difference in potential toxicity between them. You can think of D_2 (ergocalciferol) as a synthetic form, although small amounts occur naturally in mushrooms. Mushrooms are an abundant source of a cholesterol-like compound called ergosterol that can be converted to vitamin D_2 when exposed to ultraviolet light, giving sun-dried mushrooms a reputation for being rich in D_2. Vitamin D_2 in supplements is typically made from irradiating an extract of yeast. Vitamin D_3 (cholecalciferol) is the form of vitamin D made by human skin and found in animal foods.

Vitamins D_2 and D_3 were long considered equivalent because they both cure rickets. However, the carrier protein that transports vitamin D in blood has a greater affinity to bind D_3. If the vitamin isn't fully bound to a carrier protein, it is free to cause toxic symptoms. This likely explains the greater toxicity of vitamin D_2. When vitamin D toxicity occurred after intentional intake of vitamin D, it was almost exclusively the D_2 form that was used. Put another way, vitamin D_2 occasionally causes toxic symptoms at a reasonable daily dose, whereas D_3 intoxication happens only with accidental overingestion or industrial incidents.

As I mentioned in the discussion of vitamin A, the symptoms of vitamin D poisoning are very similar to the symptoms of vitamin A deficiency and vice versa. Although scientific literature states that vitamins D and A are antagonistic, the term "complementary" would be a more appropriate description. Understanding exactly how vitamin D or A poisoning produces toxic symptoms helps explain the essential connection between A, D and K_2.

Vitamin D increases the production of vitamin K_2–dependent proteins. If vitamin K_2 is lacking, vitamin D toxicity will create a functional deficiency of vitamin K_2, as all that osteocalcin and MGP go hungry. Vitamin A reduces the production of vitamin K_2–dependent proteins, so it tempers vitamin D's ability to wreak havoc with all those uncarboxylated K_2–dependent proteins.[56]

In the words of fat-soluble-vitamin expert Chris Masterjohn, "If there is one, single most important shortcoming in the research investigating the toxicity of vitamin D in humans, it is that despite decades of controlled animal experiments showing that each of the fat-soluble vitamins protects against the toxicity of the others, research in humans continues to address the toxicity of vitamin D as if its actions were independent of vitamins A, E and K."[57] The same can be said for the toxicity of vitamin A.

Getting Vitamin D from the Sun

Humans and animals synthesize vitamin D when ultraviolet light (UVB) from the sun reacts with cholesterol in our skin to make cholecalciferol, also known as vitamin D_3. Full body exposure to ultraviolet B rays, which are predominant between 10 a.m. and 2 p.m., potentially produces a maximum equivalent to a consumption of about 10,000 IU of vitamin D.⁵⁸ It's worthwhile knowing that cutaneous production of vitamin D is a self-limiting reaction. In white skin, midsummer, midday vitamin D synthesis maxes out after about 20 minutes. Since there's no net increase in production after that time, you can slap on the sunscreen after you've got your dose of D. Darker skin requires longer exposure to achieve the same yield, so if you have olive, brown, black or already tanned white skin, don't be too hasty with the SPF lotion. Beyond the 40th parallel, north or south, the sun is too weak to convert any vitamin D in the winter, and summer production is blunted as well. In North America, that encompasses the northern half of the United States and all of Canada.

Getting naked in the midday sun is not practical for most people, even those who work outdoors. A study conducted in Nebraska, a state that borders the cutoff line for optimal sun exposure, showed that outdoor workers in that relatively sunny region receive an average of only 2,800 IU of vitamin D per day from summer sun exposure.[59] This is a far cry from the almost theoretical maximum of 10,000 IU. By all means, go grab some rays on your lunch break, but don't count on this to provide a lot of vitamin D.

Dietary Sources of Vitamin D

Vitamin D from food follows a similar metabolic path as vitamin D made in skin, except absorbing D_3 from the intestines means you can get around the need for UV rays. The downside is that vitamin D is as challenging to obtain through diet as K_2, maybe more so. There are a couple of concentrated but unusual food sources of vitamin D, and it's otherwise tough to come by.

Vitamin D content of selected foods

Food	International units
Summer pork or bovine blood, 1 cup	4,000
High-vitamin cod liver oil, 1 tbsp	3,450
Indo-Pacific marlin, 3 1/2 oz	1,400
Chum salmon, 3 1/2 oz	1,300
Standard cod liver oil, 1 tbsp	1,200
Herring 3 1/2 oz	1,100
Bastard halibut (olive flounder) and fatty bluefin tuna 3 1/2 oz	720
Duck egg 3 1/2 oz	720
Grunt and rainbow trout 3 1/2 oz	600
Eel 3 1/2 oz	200–560
Red sea bream 3 1/2 oz	520

Mackerel, cooked 3 1/2 oz	345–440
Sockeye salmon, cooked 3 1/2 oz	360
Sardines, canned in oil, drained 3 1/2 oz	270
Whole milk (3.25% MF), fortified 3 1/2 oz	120
Chicken egg, 1 large	41
Cow's liver, cooked, 3 1/2 oz	30
Yogurt	0

Sourced from Masterjohn C. From Seafood to Sunshine: A New Understanding of Vitamin D Safety. *Wise Traditions.* 2006, 7(2).

The most concentrated vitamin D food might seem as strange and off-putting to some as K₂'s top-ranking dish, natto. But summer pork or bovine blood is not the product of some bizarre pagan ritual. It is the main ingredient in blood sausage, better known by the less straightforward name of black pudding or boudin noir. Since vitamin D is carried in the blood, rather than stored in any particular organ, this delicious British and French food traditionally would have provided a winter source of vitamin D when made with summer pork or bovine blood and stored.

At the very bottom of the list is vitamin D–free yogurt. It's included because many people are under the impression that dairy products are a great source of vitamin D. Even though most commercial dairy products are fortified with vitamin D by law, about one-quarter of Canadians aren't obtaining the puny recommended intake of vitamin D from foods. Recent research concludes that current food choices alone are insufficient to ensure optimal blood levels of vitamin D for many Canadians, especially in winter.[60]

The Relationship Between Vitamins D and K₂

Understanding vitamin D toxicity sheds light on the essential interconnection between vitamins D and K₂, although vitamin A is also involved.

In Chapter 1 I mentioned that vitamin D increases arterial calcification under specific circumstances. This bit of bad news doesn't fit into all the good press about vitamin D and has been largely ignored. There is also plenty of evidence to challenge that finding and to suggest that vitamin D is beneficial for heart health by reducing plaque. These contradictory facts are clarified by bringing K_2 into the picture with an example that illustrates the relationship between the two nutrients.

It turns out that vitamin D has a biphasic effect on vascular calcification.[61] That means too little and too much vitamin D are both associated with arterial plaque. Somewhere between too little and too much there's a sweet spot where calcium plaque in minimized. This puzzling effect is much easier to understand if you consider vitamins D and K_2 together.

Gamma-carboxylated MGP is the most potent inhibitor of vascular calcification. Vitamin D produces MGP, and K_2 activates it. Let's say you have some K_2 in your diet because you eat cheese. If you are very deficient in vitamin D, calcium plaque builds up because there's not enough carboxylated MGP to stop it. When you start taking vitamin D or spending time in the sun, MGP production increases. The K_2 that is available will activate MGP to reduce arterial plaques. Studies conclude that vitamin D deficiency causes heart disease and D supplements reduce arterial plaque, and they are correct.

However, if your vitamin D intake keeps increasing, at some point the pool of MGP being made will exceed the amount of K_2 available to activate it. The excess uncarboxylated MGP will lead to increased arterial plaque because of the lack of vitamin K_2. Studies conclude that higher levels of vitamin D are associated with more arterial plaque, and they too are correct. Insufficient vitamin D escalates the effects of vitamin K_2 deficiency, and a relative excess of vitamin D does the same.

Vitamin A minimizes our need for K_2 while working with K_2 to rid the body of calcium liberated from the arteries.

Are Vitamins A and D Hormones?

With all the news about vitamin D in recent years, somewhere along the way you might have heard it being described as a hormone. Indeed, officially mandated nutritional committee reports for both North America and Europe state respectively that vitamin D is a hormone and that vitamin D is more like a hormone than a vitamin.[62] The same could be said of vitamin A.

What does it mean to say that a vitamin is a hormone? A hormone is "a chemical released by a cell or a gland within the body that sends out messages that affect cells in other parts of the body." A vitamin is "an organic substance, present in minute amounts in natural foodstuffs, that are essential to normal metabolism; insufficient amounts in the diet may cause deficiency diseases."[63] Since vitamins by definition are found in food, and hormones, by definition, are made within the body, they should be mutually exclusive, with no substance overlapping these two categories. But vitamins A and D meet the criteria for both classifications, sort of.

I hate to rain on the vitamins-are-hormones parade; it certainly does sound very sexy to say that vitamins A and D are hormones. However, neither retinol nor cholecalciferol (the substances ingested when you take vitamin A and D supplements) are directly bioactive. Through a series of biochemical reactions, the body transforms those nutrients into a number of metabolites, including retinoic acid and calcitriol. It is these latter compounds that have the capacity to bind to cellular receptors and DNA, affecting the activity of our genes. That makes the end products of vitamin A and D metabolism hormones, without a doubt. In scientific literature, retinoic acid and calcitriol are often referred to as the "hormonally active forms" of vitamins A and D respectively.

Confusion arises, since even scientific journal articles will use the term "vitamin D" to refer to both cholecalciferol—the true vitamin found in food and supplements—and calcitriol, the hormonally active end-product of vitamin D metabolism. Calcitriol is technically known as 1,25-dihydroxyvitamin D or $1,25(OH)_2D$. With names like that, you can see why even scientists will just say "vitamin D" and assume you know what they are talking about. In truth, calling vitamin D (strictly meaning only cholecalciferol) a hormone is like calling a caterpillar a butterfly. It is more accurate to say that cholecalciferol is a prehormone. However, since the body readily converts vitamin D into the active hormonal form, it's not really worth making a fuss about the terminology.

Some experts have suggested that referring to vitamin D (and presumably vitamin A as well) as a hormone might mislead the public into thinking that supplementing with the vitamin is equivalent to hormone replacement therapy. Since it has been widely publicized that estrogen hormone replacement therapy (HRT) increases the risk of breast cancer and cardiovascular disease, confusion about the hormonal actions of vitamin D might cause people who would benefit from taking D to avoid doing so.[64] In reality, I have never encountered a person who confused taking vitamin D with HRT. Just in case that is an issue for anyone, let me clarify: taking vitamin D is not equivalent to HRT.

The current definitions of the terms "hormone" and "vitamin" are inadequate to cover what vitamins A and D really are. While it's true that the substances in the bottle we purchase or the food we eat are not active hormones, the body readily converts those molecules to bioactive substances that have the capacity to affect genetic activity. That makes these nutrients uniquely important among vitamins and profoundly impactful on our health, whatever you call them.

What about Vitamin E?

If you have a basic knowledge of nutrition, you might be wondering by now why this discussion of fat-soluble vitamins has barely mentioned vitamin E. Vitamin E is the generic name for a group of lipid-soluble nutrients known technically as tocopherols and tocotrienols. Alpha-tocopherol is the predominant form of vitamin E in human body tissues and nutritional supplements; gamma-tocopherol is the primary form of vitamin E in most plant seeds and in our diet.[65] You can see in the diagram below that the molecular structure of vitamin E looks a lot like that of vitamin K_2. The similarity ends there. Only limited evidence suggests that vitamin E participates in the same type of functions as the other fat-soluble vitamins. Although its primary function is to act as an antioxidant, vitamin E's emerging role in hormone production makes it the wildcard in this game.

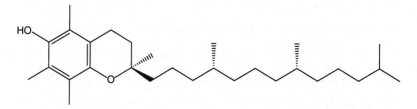

Molecular structure of vitamin E in the form of alpha-tocopherol

Vitamin E is widely considered to be the most important lipid-soluble antioxidant. Specifically, it protects against free radical damage to long-chain polyunsaturated fatty acids in the membranes of cells, which is vitally important to maintain healthy cell membranes. In fact, the more polyunsaturated fat you have in your diet—like corn, cottonseed, canola, soybean, sunflower and safflower oil—the more you need vitamin E. Commercially, alpha-tocopherol is often added to those oils to improve their shelf life and help prevent them from going rancid. Mother Nature had included gamma-tocopherol in foods that contain delicate, oxidizable polyunsaturated fats, like the yolks of grass-fed eggs and the germ of whole grains.

Whole grains are an excellent source of vitamin E if they are eaten within a day or two of grinding, since the vitamin E content drops quickly once the germ is exposed to air. Wheat germ oil contains the highest amount of vitamin E of all foods, followed by almonds and other nuts and nut butters. Most seed and grain oils (sunflower, safflower, corn) have moderate amounts of vitamin E. Trace amounts of vitamin E are found in most fruit and veggies. Avocados, with their high fat content, have more vitamin E than most fruit, but they're not exceptionally high in E. That's probably because avocados contain primarily monounsaturated fat, which is more stable and less prone to oxidation than polyunsaturated fat.

A small amount of evidence suggests that vitamin E mediates cell signaling and gene regulation, making it a hormone like vitamins A and D.[66] Other research indicates that the entire scope of vitamin E's biological activity can be understood as a function of its protection of polyunsaturated fatty acids, making antioxidant its only role.[67] If vitamin E were significantly involved in protein production, as hormones are, you would expect a deficiency of E to cause some noticeable symptoms. Instead, unlike deficiencies of A, D or K, a deficiency of vitamin E that produces clinical symptoms is rare.[68] Vitamin E has been shown to interact with the same cellular receptor as vitamins A and D, so the function of these nutrients may indeed be interrelated somehow.

Despite that vitamin E does not appear to act as a hormone directly, it does play a very important and established role in governing the release of hormones. Dietary vitamin E, or a lack of it, impacts the release of every major reproductive hormone at the level of the pituitary, the brain's master gland of hormone production. Vitamin E deficiency in animals depresses their production of follicle-stimulating hormone (FSH) and luteinizing hormone (LH), which are key fertility hormones.[69] FSH and LH control the monthly hormone cycle and ovulation in women, and sperm production in men.

Dr. Weston Price rarely mentions vitamin E in his work, but when he does it is always in the context of fertility. He centers on the fact that vitamin E is essential for healthy development of the pituitary and therefore normal production of reproductive hormones. Fertility experts in the 1930s believed that the decline in fertility that was already troubling their society was due to the reduction in vitamins B and E caused by grain milling, which had started on a large scale about one generation earlier.[70] Price referred to vitamin E as the "antisterility vitamin" and agreed with his contemporaries that diminishing vitamin E is at the root of waning female fertility in particular. And that was before we started widespread grain feeding, a practice that dramatically lowers the vitamin E content of meat and eggs. Like vitamin K₂, our modern diet is now almost devoid of vitamin E.

Price's nutritional protocol for treating dental cavities always included a whole-grain cereal, an excellent source of vitamin E. This does not mean commercial breakfast cereal as we know it today—a highly processed concoction of grains that have been puffed, flaked, shredded or generally devitalized, then "fortified." Rather, it meant porridge of whole wheat berries or other whole grain, freshly ground, soaked to remove phytic acid (an antinutrient present in bran that keeps the grain from sprouting) and cooked with milk. This provides vitamin E along with the water-soluble vitamins and the mineral cofactors to complement the A and D from cod liver oil and the K_2 from butter oil.

Vitamin E from whole foods is not a simple nutrient but a complex range of tocopherols. Grass-fed foods will restore our vitamin E intake, as will whole grains, freshly ground and properly prepared. As with most everything in nutrition, there's a heated debate about whether we should be eating grains; Chapter 8 goes there. To what extent vitamin E participates in the healing process beyond its vital role as antioxidant remains undiscovered.

No nutrient acts in a vacuum. Vitamin D, provided directly or indirectly by the sun, cooperates with A and K_2, provided more indirectly by the sun via plants and animals, to keep calcium where it should be in the body. Other nutrients are involved, too. Our mineral intake, including calcium, originates in the soil, and water-soluble vitamins naturally arrive from plant and animal foods. In the final chapter we'll take a look at where we've been, where we haven't been and where we're going.

Toward a New Definition
of Nutritious

Calcium is abundant in nature. It is the primary mineral in the sedimentary rock that covers up to 80 percent of the earth's surface, the rock that is the parent material to soil. Bones and teeth are our bodies' reservoir for calcium, holding up to 99 percent of the mineral in the human organism. Although bones have been likened to rock, really they are dynamic, living tissue that is capable of gaining and losing mineral density throughout life. Losing calcium from the skeleton compromises our health because it leads to bone fractures and opens an access route for bacteria in the mouth to reach the bloodstream. Calcium also paradoxically finds its way to places in the body that further endanger our health.

In recent years, calcium has been added to everything from multivitamins to orange juice to pasta in an effort to stave off the massive trend toward osteoporosis. Controversial research shows that this practice is, in fact, condemning calcium-takers to death from heart attack as that added calcium lodges itself in our blood vessels instead of building our bones. Simply giving up added calcium isn't the answer. Whether or not you take calcium supplements and calcium-fortified foods, it's statistically likely that hardening of the arteries, porous bones or both will affect you. That's because the problem of calcium leaching from your skeleton and gathering in your arteries is not about calcium. It is about the fat-soluble vitamins that create and activate biological proteins that guide calcium into, around and out of the body.

Even though all the fat-soluble vitamins have been known to scientists for more than 70 years, we have learned little about them until very recently. According to respected fat-soluble-vitamin researchers, this is at least partly due to financial incentive that diverts the focus of investigators toward proprietary analogs—artificial forms of vitamins that can be patented.[1] K_2 research in particular lagged behind because its sister molecule, K_1, hogged the spotlight. The fascinating menaquinone discoveries made by Dr. Weston Price remained in obscurity for decades, since K_2 goes by a pseudonym in his work. Whatever the reasons, we've got a lot of catching up to do. Here's a summary of what we know about vitamin K_2 so far:

Vitamin K_2 and our health

Health condition	Vitamin K_2 actions and benefits
Aging	• Carboxylates osteocalcin and matrix gla protein (MGP) to prevent major diseases of aging • Deficiency accelerates age-related conditions
Heart disease	• High K_2 intake lowers risk of coronary artery disease and all-cause mortality • K_2 activated–MGP is strongest inhibitor of vascular calcification presently known and prevents atherosclerosis by several mechanisms
Osteoporosis	• Activates osteocalcin, the major bone protein required for calcium deposition in bone • Deficiency increases risk of hip fracture • Counteracts bone density loss at menopause
Alzheimer's	• Protects against free radical damage and insulin resistance in the brain, two key mechanisms of brain deterioration in Alzheimer's
Wrinkles	• Lack of K_2 promotes calcification of elastic tissues in skin
Varicose veins	• K_2-activated MGP required to keep vein walls clear of calcium, just as with arteries
Diabetes	• The K_2-dependent protein osteocalcin affects insulin production and sensitivity • Supplementation improves insulin response • Higher K_2 intake associated with improved insulin sensitivity
Arthritis	• Prevents joint damage in patients with rheumatoid arthritis

Brain and neurologic health	• Shields brain cells from damage due to short-term oxygen deprivation, such as in stoke, mini-stoke or birth trauma • Required for production of myelin • Reduces severity of multiple sclerosis symptoms in animal models
Cancer	• Higher intake associated with lower rate of lung and prostate cancer • Prevents prostate cancer progression • Kills lung cancer and leukemia cells in vitro • Encourages differentiation of cancer cells
Kidney disease	• K_2 deficiency and associated blood vessel calcification increase progressively with advancing kidney disease
Fertility and pregnancy	• K_2-dependent osteocalcin affects testosterone production and sperm production and survival • Deficiency of fat-soluble vitamins associated with longer labor and higher rate of C-section
Prenatal development and children's health	• Critical for normal development of face and dental arches • Essential for normal tooth structure • Necessary for optimal growth and bone development • K_2 requirements increase during growth spurts such as puberty
Dental health	• Activates osteocalcin in tooth dentin to prevent and heal cavities • Decreases cavity-causing bacterial count in saliva

K_2 deficiency crept into our society for multiple reasons. First, the long-standing lack of awareness of the vitamin's very existence left us vulnerable to neglecting its intake. The triage theory of aging brought to light the fact that suboptimal K_2 status can go unnoticed for years before it demands our attention. Next, the gradual industrialization of our food supply that removed animals from pasture siphoned off menaquinone (K_2) in the process. This was further complicated by the introduction of trans fats and the decades-long crusade against the kinds of foods that are highest in vitamin K_2—egg yolks, cheese and butter.

Now we've turned a corner. At long last we are in a position to recognize the unique actions, deficiency symptoms and food sources of vitamin K₂. Before we conclude, let's tie up some loose ends and look at vitamin K₂ in the context of the bewildering array of popular concepts about nutrition. We now understand that vitamins K₂, A and D₃ are the foundation of health by allowing us to safely profit from all other nutrients in our diet, especially calcium. Fat-soluble-vitamin intake finally provides a litmus test for evaluating what really constitutes a healthy diet.

Defining a Healthy Diet

In the last century we have achieved unimaginable advances in the realms of science and technology, yet we still can't figure out what to have for dinner. Culture and food availability once dictated what to eat, and now that every imaginable food is available at all times, we don't know what we *should* be eating. Add to this the fact that many food cultures have influenced our own, mostly to our benefit. On the dark side, fast-food industries have notoriously waged long-standing campaigns to instill a culture around food that is profitable for them only. All this leaves the door wide open for interpretation as to what's on the menu.

The amount of conflicting nutrition advice out there can be absolutely crazy making, and I'm not even considering weight-loss diets. From the meat-heavy "ancestral" Paleolithic diet to all-plant, all-raw veganism, everybody has a theory about what to eat and some legitimate and/or pseudoscience to back it up. This book doesn't claim to solve the omnivore's dilemma but merely provides a better understanding of specific, long-misunderstood vitamins that are critical to our well-being. The true test of a nutritious diet is that it provides these vitamins, whatever the source may be. If it doesn't—because of personal food philosophy, lack of time or lack of interest—reconsider your diet or take a supplement. We don't know for sure that supplements provide all the nutrients we need in the optimal forms, doses and ratios, but we do know that they help bridge the gap.

The Paleolithic diet

If you haven't already heard of the Paleolithic diet, it may be because you haven't been living in a cave. The Paleo diet, sometimes called the caveman diet, is an eating regime based on our presumed ancestral diet. The idea behind the plan is that human genetics have not had a chance to evolve in the relatively short time since we adopted an agricultural lifestyle. Because of that, we aren't adapted to eating anything farmed or processed, so those items are off the menu. Although the concept originated in the 1970s, it has recently gained a lot of steam. Since the diet emphasizes grass-fed foods and wild game, Paleo followers were among the earliest laypeople with an awareness of vitamin K_2. The diet excludes grains, legumes, dairy, salt, refined sugar and processed oils.

A Whole Grain of Truth

Of all the modern dietary elements in the major food groups, grains are probably the most problematic. The foodstuff that for over 20 years was at the base of the recently defunct food pyramid is banished from a number of popular (like low-carb) and increasingly popular (like Paleolithic) diets. And yet Weston Price found healthy, traditional groups thriving with grains as a staple and emphasized them as an excellent source of vitamins and minerals. What gives?

This isn't as much of a detour as it might seem. Although Price successfully relied on grains to provide minerals and water-soluble vitamins when treating his patients' tooth decay, his cavity-fighting contemporaries disagreed with this approach. In particular, doctors and dental researchers Sir Edward Mellanby and Lady May Mellanby published a protocol that prevented and reversed tooth decay with a diet that was rich in fat-soluble vitamins and cereal-free.[2] They specifically asserted that the intake of whole grains neutralized the benefits of vitamin D due to the effects of a compound called phytic acid (also known as phytate).[3] If eating grains somehow defeats our vitamin D intake, which in turn jeopardizes vitamin K₂ activity, we need to get to the bottom of it.

First off, let's say we all agree that we're not considering any kind of refined grain. White flour, as cheap, convenient and tasty (in a comfort food kind of way) as it is, offers little to the nutrient balance sheet. The vitamin content, thanks to the mandatory fortification of white flour, is outweighed by the burden on our insulin production in the face of all that refined carbohydrate. We are talking whole grains. This is the unrefined cereal product that still contains the original outer bran and inner germ, along with the starchy stuff. Those in the know about phytic acid cite studies claiming that since white flour is devoid of the antinutrient, refined flour is actually better for you in some ways than whole wheat. I won't dignify that with a response, but we do need to deal with phytic acid. When it comes to grains, there are two issues: whether you are really eating what you think you are eating, and whether you should be eating it at all.

Any person with their ears tuned to the mainstream messages about healthy eating will have heard the advice about whole grain: eat more of it. Experts recommend that we eat 6 to 11 servings of whole grain every day. How much whole grain is the average, health-conscious person eating? Possibly none.

What? you ask, feeling indignant and virtuous. You have been toting that homemade sandwich on "whole-wheat" bread to lunch every day for a year. And you always choose the bread whose label says "made with whole grain" or "100 percent whole wheat." Well, what is legally allowed to pass for "whole" wheat can consist of plain old refined white flour (the bad stuff) to which had been added back as little as 30 percent of the original bran. Even dieticians consider whole-wheat flour to be merely a transitional product between completely refined flour and the actually nutritious, genuine whole grain.

Whole wheat berries, brown rice, quinoa, dehulled barley (not the standard polished, pearl barley) and whole oats are examples of legitimately complete grains. We could also lump nuts, seeds and legumes into this category insofar as all of these foods contain phytic acid. Phytic acid is a storage form of phosphorous and a compound that prevents grains from sprouting. It is the reason grains can be stored for so long if kept dry. Phytic acid accomplishes this feat by locking down minerals like zinc, iron, calcium and magnesium. Phytic acid makes these minerals unavailable for absorption. In other words, the phytic acid in whole grains defeats the purpose of eating whole grains.

Don't despair. The good news is that you can disable phytic acid by merely soaking grains (or nuts, seeds and legumes) before using them. For example, whole oats can be soaked overnight, then cooked in the morning. The traditional long rising times of sourdough bread, really a form of cultured grain, also disable the phytic acid. In contrast, quick-rising yeasted whole-wheat bread leaves phytic acid intact.

Paleo followers and the Mellanbys assert that we should deal with grains by shunning them completely. Weston Price and modern-day followers of his nutritional research assert that grains are an excellent source of essential nutrients when carefully prepared with traditional cooking techniques that disable phytic acid. Fortunately, it doesn't have to be all or nothing. Enjoy whole grains, nuts and seeds if and when you take the care to make sure they are properly prepared by soaking, sprouting or culturing, otherwise avoid them. This will safeguard your vitamin D, and by extension vitamin K_2, from the phytic acid antinutrient.

Calcium and Magnesium

With all the focus on calcium in this book, I'd be remiss if I didn't at least mention calcium's partner mineral, magnesium. Foremost among the criticisms of the calcium and heart health studies was the lack of accounting for participants' intake of magnesium. While it is K_2 that ultimately guides calcium into bones and out of soft tissue, magnesium has a profound balancing effect on calcium metabolism and is just as important as calcium for bone health. Magnesium won't reverse arterial calcification, but it offers other valuable benefits for heart health, such as lowering blood pressure. Many of the manifestations of excess or inappropriate calcification are exacerbated by magnesium deficiency. Remember from Chapter 3 the ill effect of trans fat on calcium plaque? The effect is worse when magnesium is lacking.

Aside from interacting with calcium directly, there is an essential relationship between magnesium and vitamin D that affects calcium regulation, among other things. Magnesium is essential for the absorption and metabolism of vitamin D. Magnesium deficiency, thought by many health experts to be common, impairs vitamin D metabolism. In particular, a lack of magnesium limits the conversion of vitamin D to active, hormonal form.[4] To reap the full benefits of vitamin D, and by extension vitamins K_2 and A, you need magnesium.

Magnesium-rich foods include green leafy vegetables, peas, legumes, nuts, seeds and those tricky whole grains. Be aware that long-term use of proton pump inhibitors that block stomach acid production, such as Nexium and Prevacid, deplete magnesium. Visit www .nutritionalmagnesium.org for more information about magnesium.

If you just added magnesium to your growing list of supplements for optimal bone and heart health, you might as well also jot down vitamin B_6, boron, zinc, phosphorous, vitamin C . . . If you are starting to feel a little overwhelmed, that's the point. While a good multivitamin might provide a buffer against daily dietary fluctuations, it doesn't replace good food. Now we have a starting point to define what exactly that is: food that nourishes us by providing the most important fat-soluble nutrients.

In a perfect world, the sun shines down on the backs of cows, chickens, pigs and people, creating vitamin D in the skin. Sunshine also intensifies the vitamin K_1 and beta-carotene content in grass, so grazing animals store more vitamins K_2, A and E in their fat. We humans eat that fat in a variety of delicious ways, complemented with fermented foods like cheese and natto, and the fat-soluble vitamins therein collaborate so we can profit from the minerals and water-soluble vitamins in our diet. Our children grow strong and healthy with wide smiles and straight teeth, free of decay. Our abundant nutrient intake, based on fertile, mineral-rich soil, supplies our demands not only for immediate needs but for a lifetime free of degenerative disease.

We do not need to identify every essential micronutrient and wait for scientific studies to determine their optimal daily intake to reclaim our health and the health of our children. Instead, we can let science be our informative servant and strive to follow the wise nutritional traditions of our healthy ancestors. This will be complemented, in the meantime, with supplements to provide nutrients that have become scarce in modern lifestyles.

Endnotes

Chapter 1

1. Bolland MJ, Grey A, Avenell A, et al. Calcium supplements with or without vitamin D and risk of cardiovascular events: reanalysis of the Women's Health Initiative limited access dataset and meta-analysis. *BMJ* 2011, 342:d2040.

2. Bolland MJ, Avenell A, Baron JA, et al. Effect of calcium supplements on risk of myocardial infarction and cardiovascular events: meta-analysis. *BMJ* 2010, 341:c3691; Bolland MJ, Barber PA, Doughty RN. Vascular events in healthy older women receiving calcium supplementation: randomised controlled trial. *BMJ* 2008, 336:262; Bolland MJ, Grey A, Avenall A, et al. Calcium supplements with or without vitamin D and risk of cardiovascular events: reanalysis of the Women's Health Initiative limited access dataset and meta-analysis. *BMJ* 2011, 342:d2040.

3. Magaziner J, Hawkes W, Hebel JR. Recovery from hip fracture in eight areas of function. *J Gerontol A Biol Sci Med Sci* 2000 Sep, 55(9):M498–507.

4. Sedghizadeh PP, Stanley K, Caligiuri M, et al. Oral bisphosphonate use and the prevalence of osteonecrosis of the jaw. *J Am Dent Assoc* 2009, 140(1):61–66.

5. Park-Wyllie LY, Mamdani MM, Juurlink DN, et al. Bisphosphonate use and the risk of subtrochanteric or femoral shaft fractures in older women. *JAMA* 2011, 305(8):783–89.

6. Siri-Tarino PW, Sun Q, Hu FB, et al. Saturated fat, carbohydrate, and cardiovascular disease. *Am J Clin Nutr* 2010 Mar, 91(3):502–09.

7. Geleijnse JM, Vermeer C, Grobbee DE, et. al. Dietary Intake of Menaquinone Is Associated with a Reduced Risk of Coronary Heart Disease: The Rotterdam Study. *J. Nutr* 2004, Nov 134:3100-3105.

8. Price PA, Faus SA, Williamson MK. Warfarin-induced artery calcification is accelerated by growth and vitamin D. *Arterioscler Thromb Vasc Biol* 2000 Feb, 20(2):317–27.

9. Freedman BI, Wagenknecht LE, Hairston KG. Vitamin D, adiposity, and calcified atherosclerotic plaque in African-Americans. *J Clin Endocrinol Metab* 2010, 95(3):1076–108.

10. Lee NK, Sowa H, Hinoi E, et al. Endocrine regulation of energy metabolism by the skeleton. *Cell* 2007, 130(3):456–69.

11. Oury F, Sumara G, Sumara O, et al. Endocrine regulation of male fertility by the skeleton. *Cell* 2011, 44(5):796–809.

12. Luo G, Ducy P, McKee MD, et al. Spontaneous calcification of arteries and cartilage in mice lacking matrix GLA protein. *Nature* 1997 Mar, 386(6620):78–81.

13. Schurgers LJ, Cranenburg EC, Vermeer C. Matrix gla-protein: the calcification inhibitor in need of vitamin K. *Thromb Haemost* 2008 Oct, 100(4):593–603.

14. Bolland MJ, Grey A, Avenall A, Gamble GD. Calcium supplements with or without vitamin D and risk of cardiovascular events: reanalysis of the Women's Health Initiative limited access dataset and meta-analysis. *BMJ* 2011, 342:d2040.

15. Boström K, Watson KE, Horn S, et al. Bone morphogenetic protein expression in human atherosclerotic lesions. *J Clin Invest* 1993 Apr, 91(4):1800–09.

16. Schurgers LJ, Spronk HM, Soute BA, et al. Regression of warfarin-induced medial elastocalcinosis by high intake of vitamin K in rats. *Blood* 2007 Apr 1, 109(7):2823–31.

17. Westenfeld R, Krüger T, Schlieper A, et al. Vitamin K2 supplementation reduces the elevated inactive form of the calcification

inhibitor matrix GLA protein in hemodialysis patients; Cranenburg ECM, Brandenburg VM, Vermeer C, et al. Poor vitamin K status and immature MGP species are associated with the progression of calcification in hemodialysis patients. Presented at the American Society of Nephrology Week 2008, Philadelphia.

18. Davis W. *Track Your Plaque* (New York: iUniverse, 2004), 2.

19. Storm D, Rebekah E, Smith Porter E, et al. Calcium supplementation prevents seasonal bone loss and changes in biochemical markers of bone turnover in elderly New England women: a randomized placebo-controlled trial. *J Clin Endocrinol Metab* 1998 Nov 1, 83(11):3817–25.

20. Vermeer C, Shearer MJ, Zittermann A, et al. Beyond deficiency: potential benefits of increased intakes of vitamin K for bone and vascular health. *Eur J Nutr* 2004, 43:325–35.

21. Cranenburg ECM, Schurgers LJ, Vermeer C. Vitamin K, the coagulation vitamin that became omnipotent. *Thromb Haeomost* 2007, 98(1):120–25.

Chapter 2

1. Dam H. The discovery of vitamin K. Nobel Prize. http://nobelprize .org/nobel_prizes/medicine/laureates/1943/dam-lecture.pdf.

2. Ibid., 24.

3. Hauschka PV and Reid ML. Vitamin K dependence of a calcium binding protein containing gammacarboxyglutamic acid in chicken bone. *J Biol Chem* 1978, 235:9063–68.

4. Booth SL. Skeletal functions of vitamin K–dependent proteins: not just for clotting anymore. *Nutr Rev* 1997, 55(7):282–84.

5. Cranenburg EC, Schurgers LJ, Vermeer C. Vitamin K: the coagulation vitamin that became omnipotent. *Thromb Haemost* 2007, 98(1):120–25.

6. De Oliveira C, Watt R, Hamer M. Toothbrushing, inflammation, and risk of cardiovascular disease: results from Scottish Health Survey. *BMJ* 2010, 340:c2451.

7. Price, WA. *Nutrition and Physical Degeneration*, 8th ed. (La Mesa, CA: Price-Pottenger Nutrition Foundation, 2008), 1.

8. Price WA. *Nutrition and Physical Degeneration*, 8th ed. (La Mesa, CA: Price-Pottenger Nutrition Foundation, 2008), 241.

9. Price, WA. *Nutrition and Physical Degeneration*, 8th ed. (La Mesa, CA: Price-Pottenger Nutrition Foundation, 2008), 1.

10. Lamson D. The anti-cancer effects of vitamin K. *Alt Med Review* 2003 Aug, 8(3):303-18

11. Food and Agriculture Organization of the United Nations. *Human Vitamin and Mineral Requirements*. (Bangkok, Thailand: FAO, 2002). http://www.fao.org/docrep/004/Y2809E/y2809e0g.htm

12. Geleijnse JM, Vermeer C, Grobbee DE, et al. Dietary intake of menaquinone is associated with a reduced risk of coronary heart disease: the Rotterdam Study. *J Nutr* 2004, 134:3100–05.

Chapter 3

1. Price WA. *Nutrition and Physical Degeneration*, 8th ed. (La Mesa, CA: Price-Pottenger Nutrition Foundation, 2008), 387.

2. Masterjohn C. On the trail of the elusive X-factor. *Wise Traditions* 2007, 8(1):14–32.

3. Morris ST, Purchas RW, Burnham DL. Short-term grain feeding and its effect on carcass and meat quality. Proceedings of the New Zealand Grasslands Association 1997, 57:275–77.

4. Couvreur S, Hurtaud C, Lopezet C, et al. The linear relationship between the proportion of fresh grass in the cow diet, milk fatty acid composition, and butter properties. *J Dairy Sci* 2006, 89(6):1956–69.

5. Woginrich J. Backyard chicken basics. *Mother Earth News* 2011 April/May, 245:44–48.

6. Tolan A, Robertson J, Orton CR, et al. Studies on the composition of food, the chemical composition of eggs produced under battery, deep litter and free-range conditions. *Br J Nutr* 1974, 31:185.

7. Duckett SK, Neel JPS, Fontenot JP, et al. Effects of winter stocker growth rate and finishing system on: III. Tissue proximate, fatty acid, vitamin, and cholesterol content. *J Anim Sci* 1910, doi:10.2527/jas.2009–1850.

8. Troy LM, Jacques PF, Hannan MT, et al. Dihydrophylloquinone intake is associated with low bone mineral density in men and women. *Am J Clin Nutr* 2007, 86(2):504–08.

9. Booth SL, Peterson JW, Smith D, et al. Age and dietary form of vitamin K affect menaquinone-4 concentrations in male Fischer 344 rats. *J Nutr* 2008, 138:492–96.

10. Kummerow FA, Zhou Q, Mahfouz MM. Effect of trans fatty acids on calcium influx into human arterial endothelial cells. *Am J Clin Nutr* 1999 Nov, 70(5):832–38.

11. Shurtleff W and Aoyagi A. History of natto and its relatives from history of soybeans and soyfoods: 1100 B.C. to the 1980s. Unpublished manuscript 2007. www.soyinfocenter.com

12. Kaneki M, Hedges SJ, Hosoi T, et al. Japanese fermented soybean food as the major determinant of the large geographic difference in circulating levels of vitamin K₂: possible implications for hip-fracture risk. *Nutrition* 2001, 17(4):315–21; Yaegashi Y, Onoda T, Tanno K, et al. Association of hip fracture incidence and intake of calcium, magnesium, vitamin D, and vitamin K in Japan. *Eur J Epidemiol* 2008, 23(3):219–25.

13. Ikeda Y, Iki M, Morita A, et al. Intake of fermented soybeans, natto, is associated with reduced bone loss in postmenopausal women: Japanese population-based osteoporosis. *J Nutr* 2006, 136:1323–28; Tsukamoto Y, Ichise H, Kakuda H, et al. Intake of fermented soy-bean (natto) increases circulating vitamin K₂ (menaquinone-7) and gamma-carboxylated osteocalcin concentration in normal individuals. *J Bone Miner Metab* 2000, 18(4):216–22.

14. Hsu RL, Lee KT, Wang JH, et al. Amyloid-degrading ability of nattokinase from *Bacillus subtilis* natto. *J Agric Food Chem* 2009, 57(2):503–508.

15. Ikeda Y, Iki M, Morita A, et al. Intake of fermented soybeans, natto, is associated with reduced bone loss in postmenopausal women: Japanese population-based osteoporosis. *J Nutr* 2006, 136:1323–28.

16. Vermeer C, Shearer MJ, Zittermann A, et al. Beyond deficiency: potential benefits of increased intakes of vitamin K for bone and vascular health. *Eur J Nutr* 2004, 43:325–35.

17. Van Summeren MJ, Braam LA, Lilien MR, et al. The effect of menaquinone-7 (vitamin K₂) supplementation on osteocalcin carboxylation in healthy prepubertal children. *Br J Nutr* 2009 Oct, 102(8):1171–78. Epub, 2009 May 19.

18. Geleijnse JM, Vermeer C, Grobbee DE. Dietary intake of menaquinone is associated with a reduced risk of coronary heart disease: the Rotterdam Study. *J Nutr* 2004 Nov 1, 134(11):3100–05.

19. Schurgers LJ et al. Vitamin K–containing dietary supplements: comparison of synthetic vitamin K_1 and natto-derived menaquinone-7. *Blood* 2007 Apr 15, 109(8):3279–83.

20. Sconce E, Khan T, Mason J, et al. Patients with unstable control have poorer dietary intake of vitamin K compared to patients with stable control of anticoagulation. *Thromb Haemost* 2005, 93:872–75.

Chapter 4

1. McCann JC and Ames B. Vitamin K, an example of triage theory: is micronutrient inadequacy linked to diseases of aging? *Am J Clin Nutr* 2009 Oct, 90(4):889–907, doi:10.3945/ajcn.2009.27930.

2. Ames B, et al. Low micronutrient intake may accelerate the degenerative diseases of aging through allocation of scarce micronutrients by triage. *Proc Natl Acad Sci USA* 2006, 103(47):17589–94.

3. McCann JC and Ames B. Vitamin K, an example of triage theory: is micronutrient inadequacy linked to diseases of aging? *Am J Clin Nutr* 2009 Oct, 90(4):889–907, doi:10.3945/ajcn.2009.27930.

4. Vermeer C and Theuwissen E. Vitamin K, osteoporosis and degenerative diseases of ageing. *Menopause Int* 2011, 17:19–23, doi:10.1258/mi.2011.011006.

5. McCann JC and Ames B. Vitamin K, an example of triage theory: is micronutrient inadequacy linked to diseases of aging? *Am J Clin Nutr* 2009 Oct, 90(4):889-907, doi:10.3945/ajcn.2009.27930.

6. Fonarow GC, French WJ, Frederick PD. Trends in the use of lipid-lowering medications at discharge in patients with acute myocardial infarction: 1998 to 2006. *Am Heart J* 2009 Jan, 157(1):185–94.

7. Price, WA. *Nutrition and Physical Degeneration*, 8th ed. (La Mesa, CA: Price-Pottenger Nutrition Foundation, 2008), 262.

8. Couvreur S, Hurtaud C, Lopezet C, et al. The linear relationship between the proportion of fresh grass in the cow diet, milk fatty acid composition, and butter properties. *J Dairy Sci* 2006, 89(6):1956–69.

9. Gast GC, et al. A high menaquinone intake reduces the incidence of coronary heart disease. *Nutr Metab Cardiovasc Dis* 2009 Sep, 19(7):504–10; Beulens JW, et al. High dietary menaquinone intake is associated with reduced coronary calcification. *Atherosclerosis* 2009 Apr, 203(2):489–93.

10. Geleijnse JM, Vermeer C, Grobbee DE, et al. Dietary intake of menaquinone is associated with a reduced risk of coronary heart disease: the Rotterdam Study. *J Nutr* 2004, 134:3100–05.

11. Stamler J. Diet-heart: a problematic revisit. *Am J Clin Nutr* 2010, 91:497–99.

12. Schurgers L. Regression of warfarin-induced medial elasto-calcinosis by high intake of vitamin K in rats. *Blood* 2007 Apr, 109(7):2823–31.

13. Clinical case courtesy of William Davis, MD, author of *WheatBelly: Lose the Wheat, Lose the Weight and Find Your Path Back to Health*, (New York: Rodale, 2011) and *Track Your Plaque*, 2nd ed. (New York: iUniverse, 2011).

14. Price WA. *Nutrition and Physical Degeneration*, 8th ed. (La Mesa, CA: Price-Pottenger Nutrition Foundation, 2008), 387.

15. Vehmas T, Hiltunen A, Leino-Arjas P. Seasonal variation in thoracic vessel calcifications: evidence from a chest computed tomography study. *Acta Radiol* 2010 Feb, 51(1)1:27–32.

16. Pizzorno L. Vitamin D and vitamin K team up to lower CVD risk. *Longevity Med Rev,* online reference. http://www .lmreview.com/articles/view/vitamin-d-and-vitamin-k-teamup-to-lower-cvd-risk-part-II/

17. Masterjohn C. Vitamin D toxicity redefined: vitamin K and the molecular mechanism. *Med Hypotheses* 2007, 68(5):1026–34.

18. Plaza S and Lamson D. *Alt Med Rev* 2005; Masterjohn C. *Med Hypotheses* 2007; Yamaguchi M, Sugimoto E, et al. *Mol Cell Biochem* 2001; Yamaguchi M, Uchiyama S, et al. *Mol Cell Biochem* 2003.

19. Kameda T, Miyazawa K, Mori Y, et al. Vitamin K_2 inhibits osteoclastic bone resorption by inducing osteoclast apoptosis. *Biochem Biophys Res Commun* 1996 Mar 27, 220(3):515–19.

20. Pizzorno L. Vitamin D and vitamin K team up to lower CVD risk. *Longevity Med Rev,* online reference. http://www .lmreview.com/articles/view/vitamin-d-and-vitamin-k-team-up-to-lower-cvd-risk-part-II/

21. Yamaguchi M, Uchiyama S, Tsukamoto Y, et al. Inhibitory effect of MK-7 (vitamin K2) on the bone-resorbing factors-induced bone resorption in elderly female rat femoral tissues *in vitro*. *Mol Cell Biochem* 2003, 245(1–2):115–20.

22. Kaneki M, Hedges SJ, Hosoi T, et al. Japanese fermented soybean food as the major determinant of the large geographic difference in circulating levels of vitamin K2: possible implications for hip-fracture risk. *Nutrition* 2001, 17(4):315–21.

23. Tsukamoto Y, Ichise H, Kakuda H, et al. Intake of fermented soybean (natto) increases circulating vitamin K2 (menaquinone-7) and γ-carboxylated osteocalcin concentration in normal individuals. *J Bone Miner Metab* 2000, 18(4):216–22.

24. Ramsey-Goldman R, Dunn JE, Dunlop DD, et al. Increased risk of fracture in patients receiving solid organ transplants. *J Bone Miner Res* 1999 Mar, 14(3):456–63.

25. Forli L, Bollerslev J, Simonsen S, et al. Dietary vitamin K2 supplement improves bone status after lung and heart transplantation. *Transplantation* 2010, 89(4):458–64.

26. Schurgers LJ, Teunissen KJF, Hamulyák K, et al. Vitamin K–containing dietary supplements: comparison of synthetic vitamin K1 and natto-derived menaquinone-7. *Blood* 2007, 109:3279–83.

27. Brookmeyer R, Johnson E, Ziegler-Graham K, et al. Forecasting the global burden of Alzheimer's disease. *Alzheimer's and Dementia* 2007 Jul, 3(3):186–91.

28. Loskutova N, Honea RA, Brooks WM, et al. Reduced limbic and hypothalamic volumes correlate with bone density in early Alzheimer's disease. *J Alzheimers Dis* 2010, 20(1):313–22.

29. Sparks LD. Coronary artery disease, hypertension, ApoE, and cholesterol: a link to Alzheimer's disease? *Ann NY Acad Sci* 1997, 826:128–46.

30. Presse N, Shatenstein B, Kergoat MJ, et al. Low vitamin K intakes in community-dwelling elders at an early stage of Alzheimer's disease. *J Am Diet Assoc* 2008 Dec, 108(12):2095–99.

31. Sato Y, Honda Y, Hayashida N, et al. Vitamin K deficiency and osteopenia in elderly women with Alzheimer's disease. *Arch Phys Med Rehabil* 2005 Mar, 86(3):576–81.

32. Su B, Wang X, Nunomura A, Moreira PI, Lee HG, Perry G, Smith MA, Zhu X. Oxidative stress signaling in Alzheimer's disease. *Curr Alzheimer Res* 2008, 5(6):525–32.

33. Li J, Lin JC, Wang H, et al. Novel role of vitamin K in preventing oxidative injury to developing oligodendrocytes and neurons. *J Neurosci* 2003 Jul 2, 23(13):5816–26.

34. Li J, Wang H, Rosenberg PA. Vitamin K prevents oxidative cell death by inhibiting activation of 12-lipoxygenase in developing oligodendrocytes. *J Neurosci Res* 2009 Jul, 87(9):1997–2005.

35. Li J, Lin JC, Wang H, et al. Novel role of vitamin K in preventing oxidative injury to developing oligodendrocytes and neurons. *J Neurosci* 2003 Jul 2, 23(13):5816–26.

36. Craft S. Insulin resistance syndrome and Alzheimer's disease: age- and obesity-related effects on memory, amyloid, and inflammation. *Neurobiol Aging* 2005 Dec, 26(Suppl) 1:65–69.

37. Allison AC. The possible role of vitamin K deficiency in the pathogenesis of Alzheimer's disease and augmenting the brain damage associated with cardiovascular disease. *Med Hypotheses* 2001 Aug, 57(2):151–55.

38. Pal L, Kidwai N, Glockenberg K, et al. Skin wrinkling and rigidity are predictive of bone mineral density in early postmenopausal women. *Endocr Rev* 2011, 32(03_Meeting Abstracts):3–126.

39. Park BH, Lee S, Park JW, et al. Facial wrinkles as a predictor of decreased renal function. *Nephrology* 2008, 13(6):522–27.

40. Parker BD, et al. Association of kidney function and uncarboxylated matrix gla protein: data from the Heart and Soul Study. *Nephrol Dial Transplant* 2009, 24(7):2095–101, doi:10.1093/ndt/gfp024.

41. Logan A, Levy P, Rubin MG. *Your Skin, Younger* (Naperville, IL: Cumberland House, 2010).

42. Tsukamoto Y, Ichise H, Kakuda H, et al. Intake of fermented soybean (natto) increases circulating vitamin K2 (menaquinone-7) and γ-carboxylated osteocalcin concentration in normal individuals. *J Bone Miner Metab* 2000, 18(4):216–22.

43. Geleijnse JM, Vermeer C, Grobbee DE, et al. Dietary intake of menaquinone is associated with a reduced risk of coronary heart disease: the Rotterdam Study. *J Nutr* 2004 Nov, 134(11):3100–05.

44. Gheduzzi D, Boraldi F, Annovi G, et al. Matrix gla protein is involved in elastic fiber calcification in the dermis of pseudoxanthoma elasticum patients. *Lab Invest* 2007, 87(10):998–1008.

45. Cario-Toumaniantz C, Boularan C, Schurgers LJ, et al. Identification of differentially expressed genes in human varicose veins: involvement of matrix gla protein in extracellular matrix remodeling. *J Vasc Res* 2007, 44(6):444–59.

Chapter 5

1. Lee NK, Sowa H, Hinoi E, et al. Endocrine regulation of energy metabolism by the skeleton. *Cell* 2007, 130(3):456–69.

2. Sakamoto N, Wakabayashi I, Sakamoto K. Low vitamin K intake effects on glucose tolerance in rats. *Int J Vitam Nutr Res* 1999 Jan, 69(1):27–31.

3. Sakamoto N, Nishiike T, Iguchi H, et al. Relationship between acute insulin response and vitamin K intake in healthy young male volunteers. *Diabetes Nutr Metab* 1999 Feb, 12(1):37–41.

4. Sakamoto N, Nishiike T, Iguchi H, et al. Possible effects of one week vitamin K (menaquinone-4) tablets intake on glucose

tolerance in healthy young male volunteers with different descarboxy prothrombin levels. *Clin Nutr* 2000 Aug, 19(4):259–63.

5. Iki M, Tamaki J, Fujita Y, et al. Serum undercarboxylated osteocalcin levels are inversely associated with glycemic status and insulin resistance in an elderly Japanese male population: Fujiwara-kyo Osteoporosis Risk in Men (FORMEN). *Osteoporos Int* 2011 Mar 25.

6. Hwang YC, Jeong IK, Ahn KJ, et al. The uncarboxylated form of osteocalcin is associated with improved glucose tolerance and enhanced beta-cell function in middle-aged male subjects. *Diabetes Metab Res Rev* 2009 Nov, 25(8):768–72.

7. Yoshida M, Jacques PF, Meigs JB, et al. Effect of vitamin K supplementation on insulin resistance in older men and women. *Diabetes Care* 2008 Nov, 31(11):2092–96, doi:10.2337/dc08-1204.

8. Turesson C, Jacobsson LT, Matteson EL. Cardiovascular comorbidity in rheumatic diseases. *Vasc Health Risk Manag* 2008, 4(3):605–14.

9. Morishita M, Nagashima M, Wauke K, et al. Osteoclast inhibitory effects of vitamin K2 alone or in combination with etidronate or risedronate in patients with rheumatoid arthritis: 2-year results. *J Rheumatol* 2008 Mar, 35(3):407–13.

10. Okamoto H. Vitamin K and rheumatoid arthritis. *IUBMB Life* 2008 Jun, 60(6):355–61.

11. Li J, Lin JC, Wang H, et al. Novel role of vitamin K in preventing oxidative injury to developing oligodendrocytes and neurons. *J Neurosci* 2003 Jul 2, 23(13):5816–26.

12. Thijssen HH, et al. Vitamin K status in human tissues: tissue-specific accumulation of phylloquinone and menaquinone-4. *Br J Nutr* 1996 Jan, 75(1):121–27.

13. MS Society of Canada. Genetic study supports vitamin D deficiency as an environmental factor in MS susceptibility. http://mssociety .ca/en/research/medmmo_20090205.htm

14. Moriya M, Nakatsuji Y, Okuno T, et al. Vitamin K2 ameliorates experimental autoimmune encephalomyelitis in Lewis rats. *J Neuroimmunol* 2005 Dec 30, 170(1–2):11–20.

15. Thijssen HHW and Drittij-Reijnders MJ. Vitamin K status in human tissues: tissue-specific accumulation of phylloquinone and menaquionone-4. *Br J Nutr* 1996, 75:121–27.

16. Nimptsch K, Rohrmann S, Kaaks R, et al. Dietary vitamin K intake in relation to cancer incidence and mortality: results from the Heidelberg Cohort of the European Prospective Investigation into Cancer and Nutrition (EPIC-Heidelberg). *Am J Clin Nutr* 2010, 91(5):1348–58.

17. Bostwick DG and Eble JN. *Urological Surgical Pathology* (St. Louis: Mosby, 2007), 468.

18. Nimptsch K, Rohrmann S, Linseisen J. Dietary intake of vitamin K and risk of prostate cancer in the Heidelberg cohort of the European Prospective Investigation into Cancer and Nutrition (EPIC-Heidelberg). *Am J Clinl Nutr* 2008 Apr, 87(4):985–92.

19. Nimptsch K, Rohrmann S, Nieters A, et al. Serum undercarboxylated osteocalcin as biomarker of vitamin K intake and risk of prostate cancer: a nested case-control study in the Heidelberg Cohort of the European Prospective Investigation into Cancer and Nutrition. *Cancer Epidemiol Biomarkers Prev* 2009, 18(1):49–56.

20. Yoshida T, Miyazawa K, Kasuga I. Apoptosis induction of vitamin K2 in lung carcinoma cell lines: the possibility of vitamin K2 therapy for lung cancer. *Int J Oncol* 2003 Sep, 23(3):627–32.

21. Lamson DW and Plaza SM. The anticancer effects of vitamin K. *Altern Med Rev* 2003, 8:303–18.

22. Yaguchi M, Miyazawa K, Katagiri T, et al. Vitamin K2 and its derivatives induce apoptosis in leukemia cells and enhance the effect of all-trans retinoic acid. *Leukemia* 1997, 11(6):779–87.

23. Lamson DW and Plaza SM. The anticancer effects of vitamin K. *Altern Med Rev* 2003, 8:303–18.

24. Iguchi T, Miyazawa K, Asada M, et al. Combined treatment of leukemia cells with vitamin K2 and 1alpha, 25-dihydroxy vitamin D3 enhances monocytic differentiation along with becoming resistant to apoptosis by induction of cytoplasmic p21 CIP1. *Int J Oncol* 2005 Oct, 27(4):893–900.

25. Habu D, Shiomi S, Tamori A, et al. Role of vitamin K2 in the development of hepatocellular carcinoma in women with viral cirrhosis of the liver. *JAMA* 2004 Jul 21, 292(3):358–61.

26. Otsuka M, Kato N, Shao RX, et al. Vitamin K_2 inhibits the growth and invasiveness of hepatocellular carcinoma cells via protein kinase A activation. *Hepatology* 2004, 40(1):243–25.

27. Yoshimura K, Takeuchi K, Nagasaki K, et al. Prognostic value of matrix gla protein in breast cancer. *Mol Med Report* 2009 Jul–Aug, 2(4):549–53; Levedakou EN, Strohmeyer TG, Effert PJ, et al. Expression of the matrix gla protein in urogenital malignancies. *Int J Cancer* 1992 Oct 21, 52(4): 534–37.

28. Holden RM, Morton AR, Garland JS, et al. Vitamins K and D status in stages 3–5 chronic kidney disease. *Clin J Am Soc Nephrol* 2010 Apr, 5(4):590–97.

29. Dindyal S. The sperm count has been decreasing steadily for many years in Western industrialised countries: is there an endocrine basis for this decrease? *Int J Urol* 2004, 2(1).

30. Howe AM and Webster WS. Vitamin K—its essential role in craniofacial development: a review of the literature regarding vitamin K and craniofacial development. *Austr Dent J* 1994, 39(2):88–92.

31. Price, WA. *Nutrition and Physical Degeneration*, 8th ed. (La Mesa, CA: Price-Pottenger Nutrition Foundation, 2008), 373.

32. Vieira AR and Orioli IM. Birth order and oral clefts: a meta analysis. *Teratology* 2002 Nov, 66(5):209–16.

33. Price, WA. *Nutrition and Physical Degeneration*, 8th ed. (La Mesa, CA: Price-Pottenger Nutrition Foundation, 2008), 305.

34. Price WA. *Nutrition and Physical Degeneration*, 8th ed. (La Mesa, CA: Price-Pottenger Nutrition Foundation, 2008), 305.

35. Ibid., 75.

36. Ibid., 305.

37. Merewood A, Mehta SD, Chen TC, et al. Association between vitamin D deficiency and primary cesarean section. *Journ Clin End Metab* 2009, 94(3):940–45.

38. Nishimura J, Arai N, Tohmatsu J. Measurement of serum undercarboxylated osteocalcin by ECLIA with the "Picolumi ucOC" kit. *Clin Calcium* 2007 Nov, 17(11):1702–08.

39. Van Summeren MJH, van Coeverden SC, Schurgers LJ, et al. Vitamin K status associated with childhood bone mineral content. *Br J Nutr* 2008, doi:10.1017/S0007114508921760.

40. Van Summeren MJH, Braam LA, Lilien MR. The effect of menaquinone-7 (vitamin K2) supplementation on osteocalcin

carboxylation in healthy prepubertal children. *Br J Nutr* 2009 Oct, 102(8):1171–78.

41. Price, WA. *Nutrition and Physical Degeneration*, 8th ed. (La Mesa, CA: Price-Pottenger Nutrition Foundation, 2008), 263.

42. Price WA. *Nutrition and Physical Degeneration*, 8th ed. (La Mesa, CA: Price-Pottenger Nutrition Foundation, 2008), 398.

43. Thijssen HHW, et al. Vitamin K distribution in rat tissues: dietary phylloquinone is a source of tissue menquinone-4. *Br J Nutr* 1994, 72:415–25.

44. Price, WA. *Nutrition and Physical Degeneration*, 8th ed. (La Mesa, CA: Price-Pottenger Nutrition Foundation, 2008), 263.

45. Lewis DW and Ismail AI. Prevention of dental caries. http://www .phac-aspc.gc.ca/publicat/clinic-clinique/pdf/s4c36e.pdf; Toverud G, Finn SB, Cox GJ, et al. *A Survey of the Literature of Dental Caries* (Washington, DC: National Academy of Sciences National Research Council, 1952), 165.

46. Mellanby M and Pattison CL. Remarks on the influence of a cereal free diet rich in vitamin D and calcium on dental caries in children. *Br Med J* 1932 Mar 19, 1(3715):507–10.

47. Ibid., 507.

48. Lund AE. Women have more caries than men. *J Am Dent Assoc*, 2009, 140(1):20–22.

49. Kateeb, E. Gender-specific oral health attitudes and behaviour among dental students in Palestine. *East Mediterr Health J.* 2010 Mar; 16(3):329–33.

50. Friedewald VE, Kornman KS, Beck JD, et al. The American Journal of Cardiology and Journal of Periodontology editors'

consensus: periodontitis and atherosclerotic cardiovascular disease. *J Periodontol* 2009, 80:1021–32.

51. Lalla E, Kunzel C, Burkett S, et al. Identification of unrecognized diabetes and pre-diabetes in a dental setting. *J Dent Res* 2011, 90(7):855.

52. Syrjanen J, et al. Dental infection in association with cerebral infarction in young and middle-aged men. *J Intern Med* 1989, 225:179–84; Mattila KJ, et al. Association between dental health and acute myocardial infarction. *Brit Med J* 1989, 298:779–82.

53. DeStefano F, et al. Dental disease and risk of coronary heart disease and mortality. *Brit Med J* 1993, 306:688–91.

54. Beck J, et al. Periodontal disease and cardiovascular disease. *J Periodontal* 1996, 67(Suppl):1123–37.

55. Grant WB and Boucher BJ. Are Hill's criteria for causality satisfied for vitamin D and periodontal disease? *Dermatoendocrinol* 2010 Jan, 2(1):30–36.

Chapter 6

1. Vieth R. The pharmacology of vitamin D, including fortification strategies. In *Vitamin D*, 2nd ed., ed. Feldmean D and Glorieux F (San Diego: Elsevier Academic Press, 2005) 995–1018.

2. Gundberg CM, Nieman SD, Abrams S, et al. Vitamin K status and bone health: an analysis of methods for determination of undercarboxylated osteocalcin. *J Clin Endo Metab* 1998, 83(9):3258–66.

3. Koyama N, Ohara K, Yokota H, et al. A one step sandwich enzyme immunoassay for gamma-carboxylated osteocalcin

using monoclonal antibodies. *J Immunol Methods* 1991 May 17, 139(1):17–23.

4. Hozuki T, Imai T, Tsuda E, et al. Response of serum carboxylated and undercarboxylated osteocalcin to risedronate monotherapy and combined therapy with vitamin K(2) in corticosteroid-treated patients: a pilot study. *Intern Med* 2010, 49(5): 371–76.

5. McClung MR. The relationship between bone mineral density and fracture risk. *Curr Osteoporos Rep* 2005 Jun, 3(2):57–63.

6. Iwamoto J, Takeda T, Sato Y. Role of vitamin K2 in the treatment of postmenopausal osteoporosis. *Curr Drug Saf* 2006 Jan, 1(1):87–97.

7. Jono S, et al. Matrix gla protein is associated with coronary artery calcification as assessed by electron-beam computed tomography. *Thromb Haemost* 2004, 91(4):790–94.

8. Shaw LJ, et al. Coronary artery calcium as a measure of biologic age. *Atherosclerosis* 2006 Sep, 188(1):112–19.

9. Taylor AJ, et al. Coronary calcium independently predicts incident premature coronary heart disease over measured cardiovascular risk factors: mean three-year outcomes in the Prospective Army Coronary Calcium (PACC) project. *J Am Coll Cardiol* 2005, 46(5):807–14.

10. Greenland P, LaBree L, Azen SP, et al. Coronary artery calcium score combined with Framingham score for risk prediction in asymptomatic individuals. *JAMA* 2004, 291(2):210–15.

11. Polonsky TS, McClelland RL, Jorgensen NW, et al. Coronary artery score and risk classification for coronary heart disease. *JAMA* 2010, 303(16):1610–16.

12. Elias-Smale SE, Vliegenthart Proença R, Koller MT, et al. Coronary calcium score improves classification of coronary heart disease risk in the elderly. *J Am Coll Cardiol* 2010, 56:1407–14.

13. Jono S, Ikari Y, Vermeer C, et al. Matrix gla protein is associated with coronary artery calcification as assessed by electron-beam computed tomography. *Thromb Haemost* 2004 Apr, 91(4):790–94.

14. O'Donnell CJ, Shea MK, Price PA, et al. Matrix gla protein is associated with risk factors for atherosclerosis but not with coronary artery calcification. *Arterioscler Thromb Vasc Biol* 2006 Dec, 26(12):2769–74.

Chapter 7

1. Wolf G. A history of vitamin A and retinoids. *FASEB J* 1996 Jul, 10:1102–07.

2. Ibid.

3. Price WA. *Nutrition and Physical Degeneration*, 8th ed. (Lemon Grove, CA: Price-Pottenger Nutrition Foundation, 2008), 251.

4. Masterjohn C. Vitamin A on trial: does vitamin A cause osteoporosis? *Wise Traditions* 2006, 7(1):25–41.

5. Hogart CA and Griswald MD. The key role of vitamin A in spermatogenesis. *J Clin Invest* 2010, 120(4):956–62.

6. Ortega RM, Andrés P, Martínez RM, et al. Vitamin A status during the third trimester of pregnancy in Spanish women: influence on concentrations of vitamin A in breast milk. *Am J Clin Nutr* 1997 Sep, 66(3):564–68.

7. Dalmiya N and Palmer A. *Vitamin A Supplementation: A Decade of Progress* (New York: United Nations Children's Fund, 2007), 19.

8. Formelli F, Meneghini E, Cavadini E. Plasma retinol and prognosis of postmenopausal breast cancer patients. *Cancer Epidemiol Biomarkers Prev* 2009, 18(1):42–48.

9. Logan, WS. Vitamin A and keratinization. *Arch Derm* 1972 May, 105:748–53.

10. Basu TK, Donald EA, Hargreaves JA. Seasonal variation of vitamin A (retinol) status in older men and women. *J Am Coll Nutr* 1994 Dec, 13(6):641–45.

11. Pinnock CB, Dougla RM, Badcock NR. Vitamin A status in children who are prone to respiratory tract infections. *Aust Paediatr J* 1986, 22:95–99.

12. Fallon S. Vitamin A saga. *Wise Traditions* 2001, 2(4).

13. Stephens D, Ludder Jackson P, Gutierrez Y. Subclinical vitamin A deficiency: a potentially unrecognized problem in the United States. *Pediatric Nursing* 1996 Sep–Oct, 22(5):377–89, 456.

14. Gerster H. Vitamin A—functions, dietary requirements and safety in humans. *Int J Vit Nutr Res* 1997, 67:71–90.

15. Ibid.

16. Ibid.

17. Johnson EJ, Krall EA, Dawson-Huges B, et al. The lack of effects of multivitamins containing vitamin A on serum retinyl esters and liver function tests in healthy women. *J Am Coll Nutr* 1992, 11:682–86; Sibulesky L, Hayes KC, Pronczuk A, et al. Safety of <7500 RE (<25000 IU) vitamin A daily in adults with retinitis pigmentosa. *Am J Clin Nutr* 1999, 69(4):656–63.

18. Masterjohn C. Vitamin A on trial: does vitamin A cause osteoporosis? *Wise Traditions* 2006 7(1):25–41.

19. Oliva A, Ragione FD, Fratta M, et al. Effect of retinoic acid on osteocalcin gene expression in human osteoblasts. *Biochem Biophys Res Commun* 1993, 191(3):908–14.

20. Metz AL, Walser MM, Olsen WG. The interaction of vitamins A and D related to skeletal development in the turkey poult. *J Nutr* 1985 Jul, 115(7):929–35.

21. Johansson S and Melhus H. Vitamin A antagonizes calcium response to vitamin D in man. *J Bone Miner Res* 2001 Oct, 16(10):1899–905.

22. Nau H, Chahoud I, Dencker L, et al. *Teratogenicity of vitamin A and retinoids in vitamin A in health and disease.* (New York: Marcel Dekker, 1994), 617.

23. Lelièvre-Pégorier M, Vilar J, Ferrier ML, et al. Mild vitamin A deficiency leads to inborn neprhon deficit in the rat. *Kidney Int* 1998, 54:1455–62; Chailley-Heu B, Chelly N, Lelièvre-Pégorier M, et al. Mild vitamin A deficiency delays fetal lung maturation in the rat. *Am J Respir Cell Mol Biol* 1999 Jul, 21(1):89–96.

24. Masterjohn C. Vitamin A on trial: does vitamin A cause osteoporosis? *Wise Traditions* 2006, 7(1):25–41.

25. Booth SL, Johns T, Kuhnlein H. Natural food sources of vitamin A and provitamin A: difficulties with the published values. *United Nations University Press Food and Nutrition Bulletin*, 1992 Mar, 14(1).

26. Novotny JA, Harrison DJ, Pawlosky R, et al. Beta-carotene conversion to vitamin A decreases as the dietary dose increases in humans. *J Nutr* 2010 May, 140(5):915–18.

27. Erdman J. The physiologic chemistry of carotenes in man. *Clin Nutr* 1988, 7(3):101–06. For a fascinating and more in-depth look at the

history, perception and misperception of vitamin A, read Fallon
S and Enig M. Vitamin A saga. *Wise Traditions* 2001, 2(4). http://
habitation.westonaprice.org/journal/1282-journal-winter-2001.html

28. Wood M. New clues about carotenes revealed. *Agricultural
Research* 2001 Mar, 49(3):12–13.

29. De Pee S, West CE, et al. Lack of improvement in vitamin A status with increased consumption of dark green leafy vegetables.
Lancet 1995, 346:75–81.

30. Standing Committee on the Scientific Evaluation of Dietary
Reference Intakes. *Dietary Reference Intakes: Calcium, Phosphorus,
Magnesium, Vitamin D, and Fluoride* (Washington, DC: National
Academy Press, 1997); Health and Consumer Protection Directorate-
General. Opinion of the scientific committee on food on the toler-
able upper intake level of vitamin D, European Commission. http://
ec.europa.eu/food/fs/sc/scf/out157_en.pdf

31. Robertson WG, Gallagher JC, Marshal DH, et al. Seasonal variations in urinary excretion of calcium. *BMJ* 1974, 4:436–37.

32. Basu TK, Donald EA, Hargreaves JA, et al. Seasonal variation of
vitamin A (retinol) status in older men and women. *J Am Coll Nutr*
1994 Dec, 13(6):641–45.

33. Masterjohn C. Vitamin A on trial: does vitamin A cause osteoporosis? *Wise Traditions* 2006, 7(1):25–41.

34. Melamed ML, Michos ED, Post W. 25-hydroxyvitamin D levels and
the risk of mortality in the general population. *Arch Intern Med*
2008 Aug 11, 168(15):1629–37.

35. Bischoff HA, Stahelin HB, Dick W, et al. Effects of vitamin D and
calcium supplementation on falls: a randomized controlled trial.
J Bone Miner Res 2000, 15(6):1113–18.

36. Lappe JM, Travers-Gustafson D, Davie KM. Vitamin D and calcium supplementation reduces cancer risk: results of a randomized trial. *Am J Clin Nutr* 2007 Jun, 85(6):1586–91.

37. Rajakumar K, de las Heras J, Chen TC, et al. Vitamin D status, adiposity, and lipids in black American and Caucasian children. *J Clin Endocrinol Metab* 2011, May;96(5):1560–07.

38. Alvarez JA and Ashraf A. Role of vitamin D in insulin secretion and insulin sensitivity for glucose homeostasis. *Int J Endocrinol* 2010, 2010:351385.

39. Iftekhar UM, Uwaifo GI, Nicholas WC, et al. Does vitamin D deficiency cause hypertension? Current evidence from clinical studies and potential mechanisms. *Int J Endocrinol* 2010, article ID 579640.

40. Rostand SG. Ultraviolet light may contribute to geographic and racial blood pressure differences. *Hypertension* 1997 Aug, 30(2, Pt 1):150–56.

41. Fiscella K, Winter P, Tancredi D, et al. Racial disparity in blood pressure: is vitamin D a factor? *J Gen Intern Med* 2011, doi:10.1007/s11606-011-1707-8.

42. Griffin FC, Gadegbeku CA, Sowers MFR. Vitamin D and subsequent systolic hypertension among women. *Am J Hypertens* 2011 Mar, 24:316–21.

43. Pfeifer M, Begerow B, Minne HW. Effects of a short-term vitamin D(3) and calcium supplementation on blood pressure and parathyroid hormone levels in elderly women. *J Clin Endocrinol Metab* 2001 Apr, 86(4):1633–37.

44. Embry AF, Snowdon LR, Vieth R. Vitamin D and seasonal fluctuations of gadolinium-enhancing magnetic resonance imaging lesions in multiple sclerosis. *Ann Neurol* 2000, 48:271–72.

45. Stene LC, Ulriksen J, Magnus P. Use of cod liver oil during pregnancy associated with lower risk of type I diabetes in the offspring. *Diabetologia* 2000 Sep, 43(9):1093–98.

46. Hypponen E, Laara E, Reunanen A, et al. Intake of vitamin D and risk of type 1 diabetes: a birth cohort study. *Lancet* 2001, 358(9292):1500–03.

47. Cannell JJ, et al. Epidemic influenza and vitamin D. *Epidemiol Infect* 2006, 134:1129–40.

48. Campbell LA, Kuo CC, Grayston JT. Chlamydia pneumoniae and cardiovascular disease. *Emerging Infect Dis* 1998, 4(4): 571–79.

49. Donati M, Di Leo K, Benincasa M, et al. Activity of cathelicidin peptides against chlamydia spp. *Antimicrob Agents Chemother* 2005, 49(3):1201–02.

50. Jablonski NG and Chaplin G. The evolution of human skin coloration. *J Hum Evol* 2000, 39(1):57–106.

51. Pearce SHS and Cheetham TD. Diagnosis and management of vitamin D deficiency. *BMJ* 2010, 340:b5664.

52. Vieth R. The pharmacology of vitamin D, including fortification strategies. In *Vitamin D*, 2nd ed., ed. Feldmean D and Glorieux F (San Diego: Elsevier Academic Press, 2005) 995–1018.

53. Clemens TL, Adams JS, Henderson SL, et al. Increased skin pigment reduces the capacity of skin to synthesise vitamin D3. *Lancet* 1982, 1(8263):74–76.

54. Vieth R. Critique of the considerations for establishing tolerable upper intake levels for vitamin D. *J Nutr* 2006 Apr, 136(4):1117–22.

55. Price PA, Faus SA, Williamson MK. Warfarin-induced artery calcification is accelerated by growth and vitamin D. *Arterioscler Thromb Vasc Biol* 2000 Feb, 20(2):317–27.

56. Masterjohn C. Vitamin D toxicity redefined: vitamin K and the molecular mechanism. *Med Hypotheses* 2007, 68(5):1026–34; Fu X and Wang XD. 9-Cis retinoic acid reduces 1,25-dihydroxycholecalciferol-induced renal calcification by altering vitamin K–dependent-carboxylation of matrix-carboxyglutamic acid protein in A/J male mice. *J Nutr* 2008, 138(12):2337–41.

57. Masterjohn C. From seafood to sunshine: a new understanding of vitamin D safety. *Wise Traditions* 2006, 7(2). http://www.westonaprice.org/fat-soluble-activators/from-seafood-to-sunshine-a-new-understanding-of-vitamin-d-safety

58. Vieth R. The pharmacology of vitamin D, including fortification strategies. In *Vitamin D*, 2nd ed., ed. Feldmean D and Glorieux F (San Diego: Elsevier Academic Press, 2005) 995–1018.

59. Barger-Lux MJ and Heaney RP. Effects of above average summer sun exposure on serum 25-hydroxyvitamin D and calcium absorption. *J Clin Endocrinol Metab* 87(11):4952–56.

60. Whiting SJ, et al. The vitamin D status of Canadians relative to the 2011 Dietary Reference Intakes: an examination in children and adults with and without supplement use. *Am J Clin Nutr* 2011 Jul, 94(1):128–35.

61. Zittermann A, Schleithoff SS, Koerfer R. Vitamin D and vascular calcification. *Curr Opin Lipidol* 2007 Feb, 18(1):41–46.

62. Vieth R. Why "vitamin D" is not a hormone, and not a synonym for 1,25-dihydroxy-vitamin D, its analogs or deltanoids. *J Biochem Mol Bio* 2004, 89–90:571–73.

63. *Stedman's Medical Dictionary*, 27th edition, s.v. "hormone" and "vitamin" (Baltimore: Lippincott Williams and Wilkins, 2000).

64. Fu X and Wang XD. 9-Cis retinoic acid reduces 1,25-dihydroxy-cholecalciferol-induced renal calcification by altering vitamin K–dependent-carboxylation of matrix-carboxyglutamic acid protein in A/J male mice. *J Nutr* 2008, 138(12):2337–41.

65. Jiang Q, Christen S, Shigenaga MK, et al. Gamma tocopherol, the major form of vitamin E in the US diet, deserves more attention. *Am J Clin Nutr* 2001, 74(6):714–22.

66. Azzi A, Gysin R, Kempna P, et al. Vitamin E mediates cell signaling and regulation of gene expression. *Ann NY Acad Sci* 2004, 1031:86–95.

67. Traber MG and Atkinson J. Vitamin E, antioxidant and nothing more. *Free Radic Biol Med* 2007, 43(1):4–15.

68. Institute of Medicine, Food and Nutrition Board. *Dietary Reference Intakes: Vitamin C, Vitamin E, Selenium, and Carotenoids* (Washington, DC: National Academy Press, 2000).

69. Akazawa N, Mikami S, Kimura S. Effects of vitamin E deficiency on the hormone secretion of the pituitary-gonadal axis of the rat. *Tohoku J Exp Med* 1987 Jul, 152(3):221–29.

70. Drummond JC. The medical aspects of decline in populations. *JAMA* 1938, 110:908.

Chapter 8

1. Vieth R. The pharmacology of vitamin D, including fortification strategies. In *Vitamin D*, 2nd ed., Feldman D and Glorieux F (San Diego: Elsevier Academic Press, 2005) 995–1018.

2. Mellanby M and Pattison CL. Remarks on the influence of a cereal free diet rich in vitamin D and calcium on dental caries in children. *Br Med J* 1932 Mar 19, 1(3715):507–10.

3. Mellanby E. The rickets-producing and anti-calcifying action of phytate. *J Physiol* 1949 Sep 15, 109(3–4):488–533.

4. Zofková I and Kancheva RL. The relationship between magnesium and calciotropic hormones. *Magnes Res* 1995 Mar, 8(1):77–84.

Index